Reptiles of the
Australian
High Country

A FIELD GUIDE TO

Reptiles of the Australian High Country

Robert Jenkins
and
Roger Bartell

INKATA PRESS

INKATA PRESS PROPRIETARY LIMITED
MELBOURNE

PUBLISHED 1980
©R. W. G. Jenkins and R. J. Bartell

National Library of Australia card
number and ISBN 0 909605 16 5

National Library of Australia
Cataloguing in Publication data

Jenkins, Robert W. G.
    A field guide to reptiles of the
    Australian high country.

    Index
    Bibliography
    ISBN 0 909605 16 5

    1. Reptiles — Australia — Identification.
    I. Bartell, Roger J., joint author.
    II. Title.

    598.1'0994

Set in 10/11 Baskerville by Savage & Co., Brisbane
and printed in Hong Kong

Foreword

The majority of Australians live in the south-eastern corner of the continent — a region which encompasses a remarkable diversity of landscapes, including the Australian alps.

In this book, the authors not only summarise our present knowledge of the rich and diverse reptile fauna of this region, but present many new observations. Their knowledge of, and enthusiasm for, the animals about which they write is evident throughout their work, and I have no doubt that this book will give a strong impetus to the study and conservation of this much-maligned group of animals.

Reptiles, including snakes, are now given statutory protection in most parts of Australia — no mean achievement in a country which has traditionally held these animals in low regard. However, there is always the danger that the statutory protection of species may be used as a substitute for conserving adequately those ecosystems or communities of which the species are an integral part, and on whose survival that of the species will ultimately depend. In the race to produce more timber, or uranium, or wheat, these ecosystems are under mounting pressure in Australia as in other parts of the world; their survival may prove to be the measure of man's ability to come to grips with the ecological and sociological problems which now threaten the very survival of a large proportion of our planet's biota.

This book will be welcomed by both specialist and layman for its use in helping to identify those reptiles found in a populous area of south-eastern Australia. Perhaps it is more important that it will help Australians to gain a better understanding of the place of reptiles in natural systems, and so increase the knowledge and awareness of man's place in those systems.

Harold G. Cogger
Curator of Reptiles and Amphibia
The Australian Museum
Sydney   *February 1979*

Acknowledgements

The authors are grateful to the many colleagues and friends who, directly or indirectly, have contributed to the preparation of this book. In particular, they wish to thank the following persons, without whose generous help the preparation of the manuscript would have taken considerably longer: Mr Ian Morris, for photographs of *Austrelaps superba, Pseudonaja textilis, Notechis scutatus, Acanthophis antarcticus, Hemiergis maccoyi, Egernia whitii* and *Leiolopisma trilineata*; Mr John Wombey for photographs of *Typhlina nigrescens, Amphibolurus muricatus* and *Chelodina longicollis*; Dr Harold Cogger for photographs of *Menetia greyii, Furina diadema* and *Vermicella annulata*; Mr Frank Collet for the photograph of *Tympanocryptis lineata*; Doctors Barwick, Cogger and Moore, and Messrs Longmore and Wombey for reviewing the manuscript; Doctors Cogger, Greer, Rawlinson and Storr for taxonomic advice; Ms Sybil Monteith for illustrations of the head scalation of *Tiliqua* and *Typhlina*, and of the caudal regions of *Unechis gouldii* and *Pseudonaja textilis*. Thanks are also given to: Mr J. Coventry of the National Museum of Victoria for assistance with field work; Mr J. Green of CSIRO for technical assistance in the selection of photographs for colour reproduction in the plates, and Mr C. Totterdell, also of CSIRO, for the preparation of working colour proofs; and Mrs M. Hawke for typing of the taxonomic working drafts. We are also indebted to the Directors of the National Museum of Victoria, the Australian Museum in Sydney, and to Dr H. Frith of CSIRO Division of Wildlife Research (Australian National Wildlife Collection) for allowing us access to museum specimens and records. Finally, and by no means least, thanks to our wives and families who have had to put up with our absences while collecting in the field, our periodic seclusion while writing, and who have from time to time imperturbably given houseroom to miscellaneous snakes and lizards.

Contents

Introduction

In this book we have endeavoured to provide an introduction to the reptiles as an intriguing group of animals and at the same time give a comprehensive treatment of those species which inhabit a small corner of the Australian continent. Our aim has been to cater not only for those already familiar with the reptiles but also for the newcomer to the field, the bushwalker or the occasional visitor to the countryside, who simply wants to know what it is he has seen.

We are concerned that, because of the lessening ties of modern society with the natural environment (mainly through its dependence upon material prosperity), much of Australia's natural heritage could be in danger of being destroyed. It is perhaps understandable that the pioneering spirit, wherein man went out to tame the bush, should linger on within present day attitudes; some still regard the bush as an antagonist to be burned and cleared and fail to acknowledge the many things of value that it can provide. It is heartening then to find that more and more people are beginning to realise that the world of nature does have much to offer — a place for solitude, relaxation, inspiration and learning. Many are seeking to discover for themselves more about Australia's unique flora and fauna. This is demonstrated by the recent rapid increase in the number of popular wild-life books, many of which are well endowed with colour pictures and accounts of plants and animals. Few, however, go beyond the superficial treatment and there is a wide gap between this level and the detailed textbook; in writing for this intermediate level, we hope partially to remedy the imbalance.

It has taken more than six years to compile all the information presented in this book, and much of that time has been spent in securing live specimens for the photographic plates. Many of the animals proved to be exasperating and often infuriating subjects for the camera. We have attempted to show each animal in a natural pose against a background which is typical of its normal habitat, while at the same time showing the distinguishing characters of the species. We found that it was easier to pose the smaller skinks on a piece of timber up which they could climb to the end, and where they would usually stay to be photographed. Many of the larger, faster and, in the case of snakes, more pugnacious species were generally cooled down in a refrigerator before being photographed, although even then we found that we had to work fast.

The descriptive information has been drawn from published scientific works and from examination of museum specimens. Many of the locality records have been established by finding road casualty specimens or from specimens caught or observed in the area by ourselves, friends and colleagues.

We hope that what follows will prove to be interesting and instructive to the researcher and the layman alike.

R.W.G.J.   R.J.B.

Canberra 1979

1
The Southern Highlands Region

Physiographic features

In attempting to define some natural region for discussion of its native fauna, there is always the problem of recognising precise boundaries. Naturally it is to be hoped that the selected area will contain a large proportion of species which are not represented immediately outside its boundaries. Inevitably, there will be a number of species with a distribution range which overlaps the selected area and, although they may be only marginally represented, these must be included in the local record.

To some extent the boundaries we have chosen for consideration are rather arbitrary, but for the most part they correspond with the 500-metre contour line. Although we have tried to define a natural region in terms of topographic, climatic and faunistic factors, the northern boundary is somewhat artificial.

To the east of the Southern Highlands area, the boundary follows the coastal escarpment of the Great Dividing Range. This eastern rampart stretches from about Bungonia in the north to the Victorian border near Nungatta in the south. Although this represents a fairly discreet boundary, the intrusion of the sandstone system just east of Nerriga (hatched area on the map on page 8) supports typically coastal elements in the reptile fauna. Such species as the Eastern Water Dragon, Master's Snake and others which occur in this locality have been excluded from consideration in this book.

In the New South Wales part of the region, the west is bounded by the so-called 'western slopes' which stretch from about Jinjellic in the south, skirting Gundagai to the east, then turning north-eastwards to Binalong. The remainder of the northern boundary forms a convenient delineation between the Southern Highlands region and the New South Wales central uplands.

The Victorian part of the region is entirely bounded by the 500-metre contour. In the north-eastern part of the State, it runs from the border with New South Wales near Wroxham in a more or less westerly direction to Warburton before swinging north to Yea, then north-eastward back to the border.

In all, these boundaries enclose some 68 000 square kilometres of territory; the greater area (over 39 000 square kilometres) is in New South Wales.

In New South Wales, much of the area is formed on an elevated plateau from which arise numerous ranges of low hills and mountains. The floor of the plateau is gently undulating at around 650 metres above sea level and little of the region falls below that elevation, while much of it is above 1000 metres. The western ranges rise to about 2000 metres above sea level or more, with Mt Kosciusko the highest point at 2228 metres. On the plateau itself are many isolated, low hills which rise 100 to 300 metres above the surrounding terrain.

The Victorian high country is rather different in its general topography, with slopes graded to the central massif with very little of the tableland country that is represented in the New South Wales portion of the region. In the western and central sectors, the country is similar to that in the Mt Kosciusko region, while in the south west, where the elevations are not so great, many of the peaks do not extend above the tree-line.

The whole area is climatically rather varied, but generally tends to extremes. The winters are cold and frosts are commonly experienced throughout the region; the summers tend to be warm to hot. Annual and daily temperature ranges are wide. For an *ectothermic* (cold-blooded) group such as the reptiles, these extreme temperature ranges pose some rather special problems, as are described on pages 36 and 37.

The average annual rainfall may be as high as 1500 millimetres in some of the more mountainous areas and as low as 400 millimetres elsewhere. The average annual rainfall for Canberra City is about 625 millimetres, although towards the western suburbs and ranges, it can be up to 875 millimetres. There is considerable year-to-year variability in rainfall pattern, but such fluctuations probably have little direct effect on the composition of the local reptile fauna, although species which rely upon the persistence of free water might suffer temporary setbacks in drought years.

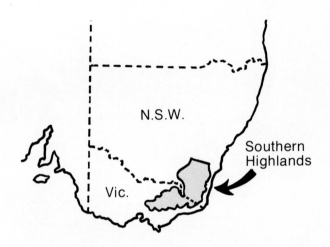

N.S.W.

Southern
Highlands

Vic.

8

9

Elevation
metres

	2000
	1500
	1000
	500
	sea level

Scale

0 10 20 30 40 50 60 70 80 90 100
kilometres

N.S.W.
VICTORIA

DELEGATE RIVER
Bendoc
Bonang

Corryong
Berringama
Talangata

Corryong

Coopracambra

Tubbut

Goongerah
Gelantipy

Murrindal

Buchan

ORBOST

LAKES ENTRANCE

BAIRNSDALE

MAFFRA

Benambra

Omeo

Ensay

Mitta Mitta

Myrtleford

Dargo

Strathbogie Range

Mansfield

OALLANDS

WARBURTON

Tooligie

Mt. Shepard

Powelltown

Neerim

Noojee

Loch

Walhalla

Woods Point

Aberfeldy

BW

Reptile habitats

For many reptile species it is difficult to be precise in identifying habitat preferences. Clearly a number of widespread species, which are also represented outside the Southern Highlands region, are to be found in a wide range of habitat types; others seem to be extremely restricted in their choice of habitat to the extent that their distribution throughout the region is rather patchy. It is evident that for many species the local climate, particularly the temperature, plays a decisive role in limiting the extent of their distributions. The prime ecological requirement for a reptile is an area which provides sufficient thermal levels to sustain activity throughout a reasonable period of the day, and through a long enough part of the year to allow time for completion of the breeding cycle, including incubation of the eggs (see Reptiles at Large, page 33). For species which inhabit forest areas this can be a severe limitation, particularly where the tree canopy is dense and closed (the leafy crowns of adjacent trees touch) and radiant heat may only be able to penetrate when the sun is at or near its zenith. No species seem to be limited by dietary requirements as most are broadly omnivorous and take a wide variety of animal and plant food.

In an endeavour to reach some uniformity of approach, we have based our treatment of habitat types as nearly as possible on that used in *Birds of the High Country*, edited by H. J. Frith. Although there are some obvious points of difference, we feel that the classification used in that book is suited to our needs.

We would like to add that the use of the classification which follows is meant only as a general guide to the distribution of reptiles and does not imply any specific habitat preference on the part of each species.

The breakdown into separate vegetational and land-form types will enable the reader to anticipate which species of reptile are likely to be encountered in a given area.

10

Broadly speaking, the region may be sub-divided into three component land-forms; the **lowland plains**, the **mountain slopes and gullies** and the **mountain ridges**.

Lowland plains

In this habitat classification, the area includes the grasslands, open woodland savannahs, lowland swamps and rivers. The area is typified by gently undulating country which is elevated little above 650 metres, and from which rise many isolated low hills. Most of the vegetation on these plains has been considerably disturbed and modified since settlement, due to clearing, burning and grazing pressures. Fortunately some areas still remain free from the intrusions of man and his domestic animals. These areas provide us with some insights into the character of the original environment, while also acting as a refuge for reptiles which might otherwise have vanished from the region. On the other hand, the clearing of land to pastures has certainly resulted in the extension of the ranges of species inhabiting the open grasslands. In this process, the common practice of ring-barking trees then leaving them to weather and fall may have contributed to the spread of these grassland species by providing abundant refuges for them.

(i) Grasslands

The floors of the valleys at around 650 metres were originally dominated by the native Kangaroo Grass *Themeda australis*. This grass and others, such as *Poa* and *Danthonia* species, have been progressively displaced since settlement by exotic grasses and herbs which have been introduced by graziers for pasture improvement. Also, in association with European man's intrusion into the area, many common weeds and garden escapees have become established.

Thirty species of reptile have been recorded from these grassland plains. Some of these are exceedingly abundant, like the Grass Skink *Leiolopisma delicata*, while others, like *Menetia greyii* and the Stone Gecko *Diplodactylus vittatus*, are patchy in

11

their distributions and occur only where refuge cover is provided by rocks and fallen timber.

(ii) Savannah woodlands

At a slightly lower elevation than the typical dry sclerophyll forests are the open savannah woodlands. The eucalypts are usually well proportioned, widely spaced and grow to a height of around 30 metres. The most common species of eucalypt in this zone are *Eucalyptus rossii, E. mannifera, E. melliodora, E. bridgesiana, E. polyanthemos, E. blakelyi* and *E. macrorhyncha*.

Since human settlement of the area, much of this zone has been considerably modified and the ground cover between trees has been degraded or replaced by exotic grasses and herbs. It is quite likely that much of the area now described under this sub-heading was formerly occupied by dry sclerophyll forest.

The reptile fauna of this zone is very rich and, as can be seen in the table on page 30, there are some 33 species listed

as occurring there. Certain of these are common to the dry sclerophyll forests and probably represent relict populations from the former climax vegetation. However, other groups, like the pygopodids, may have penetrated from the grasslands as the tree cover declined and the understorey and ground cover changed over from ericoid shrubs to grasses.

(iii) Lowland swamps

There are few of these lowland swampy areas on the plateau, the most notable and extensive being at the southern end of Lake George near the township of Bungendore. These swamps are typically vegetated by sedges and rushes and are bordered by tussocky grasses.

Sixteen species of reptile inhabit these permanent or semi-permanent lowland swamps. It is in such places that the Long-necked Tortoise *Chelodina longicollis* is most commonly to be found. The swamps also are frequented by the Mainland Tiger Snake *Notechis scutatus* where large numbers of this species occur.

(iv) Lowland rivers

Several major rivers of inland Australia have their headwaters in the ranges to the west of the region. The upper reaches of the larger rivers and streams consist of numerous, fast-flowing and often precipitous creeks. The habitats bordering the latter are categorised under 'Mountain slopes and gullies'.

The lower and slower-flowing rivers support a rich reptile fauna along their margins and 29 species have been recorded. Common amongst these are the Gippsland Water Dragon *Physignathus lesueurii howittii*, the Striped Skink *Ctenotus robustus*, the Water Skink *Sphenomorphus tympanum* (Warm Temperate Form), and the Jacky Lizard *Amphibolurus muricatus*.

Mountain slopes and gullies

This sub-division includes the dry and intermediate sclerophyll forests, and upland swamps and rivers with patches of wet sclerophyll vegetation in some of the wetter and more protected gullies of the more northern parts of the region, to the extensive wet sclerophyll forests in the Victorian uplands.

Altitudes range from above 650 metres to around 1300 metres. The rainfall is higher than on the lowland plains, and in winter the precipitation commonly falls as snow above 1000 metres.

Also included in this sub-division are the many scattered, low hills (referred to on page 16) of which very few have escaped the usually devastating impact of man. Unlike the soils of the valley floors, the soils of the hills only shallowly cover the bed-rock, and there are many rocky outcrops on the hillsides. Because of the thinness of these soils, the removal of vegetation by clearing, grazing, trampling or other interference inevitably results in erosion and accelerated destruction of the plant and animal communities supported there.

(i) Dry sclerophyll forest

This type of forest is typical of the lower hills throughout the plateau. Unfortunately there is little that has not suffered from the impact of man in some way, and there are few examples of this type of forest which support the full diversity of plant and animal species typical of the vegetational climax. Hence, most of what remains is a patchwork of secondary regeneration which is often deficient in one or more of the indicator eucalypt species. The area under dry sclerophyll forest is continually shrinking because of agriculture, burning, logging and pasture establishment. It is also the most vulnerable vegetational type for the establishment of pine plantations.

One particularly rich assemblage is to be found on Black Mountain adjacent to the central area of Canberra. Although the area is technically a reserve there have been, and continue to be, considerable intrusions which menace this fine example of dry sclerophyll forest.

The dominant trees, *Eucalyptus macrorhyncha, E. rossii, E. dives, E. sieberi* and *E. mannifera* spp. *maculosa*, rarely grow to more than 20 metres in height. The canopy is characteristically semi-closed and beneath it there is a dense understorey of low-growing and frequently prickly, ericoid shrubs of 1 to 2 metres or less in height. The richness and diversity of this vegetation is reflected in the abundance of the reptile fauna and about 32 species are to be found in this zone.

(ii) Intermediate sclerophyll forest

This vegetational sub-division has been adopted to include sclerophyll forest types which do not readily fit into the existing 'wet' or 'dry' categories. The dominant eucalypt species are *Eucalyptus sieberi* are *E. dalrympleana*, *E. viminalis* and *E. pauciflora* which grow where the annual rainfall is about 890 millimetres. Tree height ranges from 10 to 20 metres according to species and their relative exposures to weathering. Where they are fully exposed to the north west, the trees (*Eucalyptus radiata*, *E. pauciflora*, *E. fraxinoides*) may take on a maleeform habit and become dwarfed to around 1.5 metres in height. Ground cover throughout the forest is uniform and is dominated by the grasses *Poa* and *Danthonia pallida*. There are few shrubs of which *Acacia dealbata* is the most representative.

As might be expected from the intermediate nature of this vegetation type between dry sclerophyll forest on the one hand and the wet sclerophyll on the other, the reptiles to be found inhabiting this zone are also common to either one or both of the latter two vegetational types. Nineteen species of reptile are represented.

(iii)   Wet sclerophyll forest

The distribution and extent of this vegetational type in the region is rather varied. In the northern areas, and for the most part, this type of plant community is restricted to the more protected gullies of the mountain slopes. While in East Gippsland particularly, this type of forest is the most extensive vegetational form.

The zone is typified by a tall, dense growth of the dominant eucalypts (*Eucalyptus fastigata, E. robertsonii, E. viminalis*) which form a closed canopy. Beneath this canopy the Blackwood Wattles, *Acacia melanoxylon*, form a secondary tree layer. Below this again is a dense understorey of ferns and broad-leaved shrubs which grow up to 2 metres high. These include species of *Bedfordia, Drimys, Hakea, Lomatia, Olearia, Persoonia, Pomaderris, Tieghmopanax* and others. There are also many climbing plant species including *Clematis* and *Smilax*.

The gullies support a thick vegetation of ferns and tree ferns, and the logs and stones are thickly invested with mosses and liverworts. This structured wet sclerophyll forest prevents a significant proportion of solar radiation from penetrating to the forest floor, except in natural clearings around swamps, along watercourses or rocky outcrops. These forests are then largely inaccessible to the majority of reptile species which rely on sun-basking to gain body heat.

This is reflected in the observation that only nine species of reptile are to be found in the wet sclerophyll zone. The most notable of these for their abundance are *Hemiergis maccoyi*, both forms of *Sphenomorphus tympanum*, and *Leiolopisma coventryi*. Most other species listed in the table are considered to be only marginally represented in the zone.

(iv) Upland swamps and watercourses

The ranges of mountains are dissected by numerous small creeks and streams. In the higher parts, these flow through the deep gullies which are typically vegetated with wet or intermediate sclerophyll forest. In the lower areas, however, these streams flow more openly through the dry sclerophyll forests and support a characteristic reptile fauna along their margins.

Commonest among the 15 species found in these habitats are the two forms of the Water Skink *Sphenomorphus tympanum* and, especially below 1000 metres, the Gippsland Water Dragon *Physignathus l. howittii.*

Where the water-table closely underlies the surface, particularly in the more shallow valleys and hill folds, there may often be areas of swamp or marshy ground which are typified by their vegetation of sedges, tussocky grasses, ferns, mosses and liverworts. Common reptile inhabitants of these areas are *Hemiergis decresiensis, Leiolopisma trilineata,* and both forms of the Water Skink *Sphenomorphus tympanum.*

Mountain ridges

The ridges above 1300 metres support the sub-alpine and alpine communities ranging from forests to open meadows. In the winter months, snow is frequent, remaining as drifts in hollows throughout this season and persisting often into early spring. The snow-line is designated as 1500 metres above sea level, but precipitation as snow may often occur down to 1000 metres. Precipitation levels are generally above 1550 millimetres per year. In summer, the day temperatures may rise to 25°C but nights are always cool.

The ridges are characteristically steep and they are invested with shallow, poor soils. There are numerous rocky outcrops ranging from high cliffs with loose screes below to scattered heaps of boulders. The shallow valleys between the ridges are usually well vegetated although drainage is poor and peat bogs are common.

The reptile species inhabiting these areas remain in their winter retreats until the warmer seasons are well advanced. Most species construct deep burrows beneath rocks or within the loose screes. Two skink species, *Pseudemoia spenceri* and *Leiolopisma entrecasteauxii* (Form B), inhabit the deep fissures in weathered fallen timber. Forestry management practices in thinning woodland understoreys and the clearing of ski-slopes has led to an increase in the availability of refuges for these species.

(i) Sub-alpine woodlands

This zone is dominated by two species of low-growing eucalypt, *Eucalyptus niphophila* and *E. pauciflora*. The understorey
vegetation ranges from small shrubs (less than 1.5 metres high)
to herbs and grasses.

There are some 16 species of reptile inhabiting this zone.

(ii)   Alpine communities

There are several types of alpine community which exist above the timber-line. These may be recognised as tussock grasslands, herbfields, heaths and bogs.

The reptile fauna of these high areas is poorly represented by species, and those that do exist there are not particularly restricted to any one type of plant community. What the zone lacks in numbers of species, however, is well compensated for by the numbers of individuals. By far the most abundant is Form B of *Leiolopisma entrecasteauxii*. The wetter habitats are frequented by two species of water skink, *Sphenomorphus tympanum* (Cool Temperate Form) and *S. kosciuskoi*. In some localities, these two species are sympatric.

We have summarised the reptile fauna within each of the foregoing habitat categories in the accompanying table (page 30) and diagram (page 28). Additional reference to habitat preference and localities is made in the species-by-species treatment in the latter part of this book.

Other (man-made) habitats

Before concluding this chapter, we should consider two other habitat types, both of which are man-made.

Pine plantations

Since the early part of this century, it has been common forestry practice to plant extensive areas, much of which formerly supported dry sclerophyll forest, with exotic softwoods, mainly *Pinus radiata*. The consequence of widespread clearing of native forests has been the destruction of the habitats of those species of reptile which originally inhabited the dry sclerophyll forests. Although some species may still be found within the pine plantations, careful observation will show that they are usually only persisting in areas of natural bushland which have either escaped the clearing or subsequently intruded along watercourses and roadsides.

Perhaps the most tenacious reptile species in this regard are the small Grass Skink *Leiolopisma guichenoti*, the Water Skink *Sphenomorphus tympanum* (Warm Temperate Form), and *Hemiergis maccoyi*.

Suburban gardens

At more or less the same time as the exploitation of the natural bushland for pine plantation has come the explosive growth of the Federal capital, Canberra. Together with the rapid expansion of other suburban areas in the region, this particular type of land use has had considerable impact on the local environment. Not the least of these impacts has been the resumption of land for housing and supporting services. Fortunately, many people take a delight in home gardening and, in this way, provide refuges for many species of animal.

Naturally it is to be expected that those householders whose gardens border on the bush will encounter quite a few reptile species which are only casual intruders into the developed property. There are some species, however, which will take up residence and even breed in the home garden. Notable amongst these species are the Common Bluetongue *Tiliqua scincoides*, and the Grass Skink *Leilopisma guichenoti*. It is remarkable that these species manage to sustain themselves in such an apparently hostile environment, as the smaller ones must inevitably suffer considerable depredation by domestic cats.

Another frequent, but perhaps less welcome, intruder on to the domestic scene is the Common Brown Snake *Pseudonaja textilis*. Large numbers of this species tend to move into suburban areas as the summer months progress, presumably in search of a sustained food supply in the irrigated garden environment as supplies dwindle in the drier local bush. Here many of them meet an untimely end as householders fear for the safety of their pets, children and themselves — not necessarily in that order!

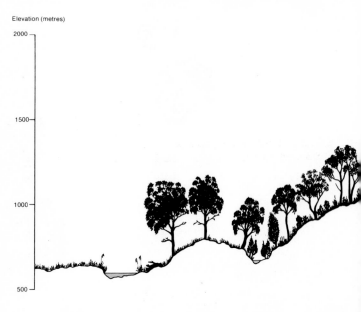

Elevation (metres)

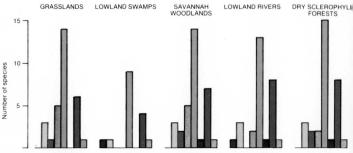

Habitat distribution of reptile families in the Southern Highlands region

The coloured vertical bars represent the numbers of species within each family occurring in each of the habitat vegetational zones. The elevation scale is provided only to indicate the approximate distribution of each vegetational zone throughout the region.

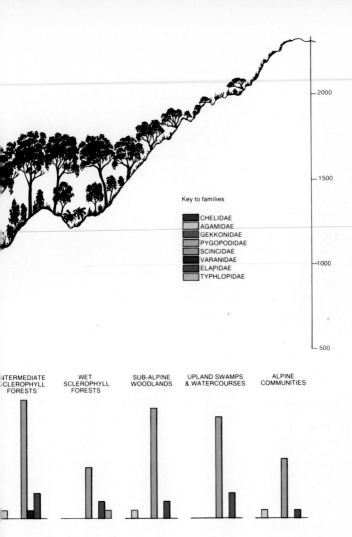

Key to families

■ CHELIDAE
□ AGAMIDAE
▨ GEKKONIDAE
▨ PYGOPODIDAE
▨ SCINCIDAE
■ VARANIDAE
■ ELAPIDAE
▨ TYPHLOPIDAE

2000

1500

1000

500

INTERMEDIATE
SCLEROPHYLL
FORESTS

WET
SCLEROPHYLL
FORESTS

SUB-ALPINE
WOODLANDS

UPLAND SWAMPS
& WATERCOURSES

ALPINE
COMMUNITIES

29

Habitat distributions of the Australian Southern Highlands Reptiles

The plus sign indicates the presence of a species in a given habitat type; a plus sign in parentheses indicates only a marginal presence.

	Grasslands	Savannah woodlands	Lowland swamps	Lowland rivers	Dry sclerophyll forests	Intermediate sclerophyll forests	Wet sclerophyll forests	Upland swamps and watercourses	Sub-alpine woodlands	Alpine communities
CHELIDAE										
Chelodina longicollis			+	+						
AGAMIDAE										
Amphibolurus barbatus	+	+	(+)	(+)	+					
A. diemensis						(+)	+		+	+
A. muricatus	+	+		+	+					
Physignathus l. howittii				+						
Tympanocryptis lineata	+	+								
GEKKONIDAE										
Diplodactylus vittatus	+	+			+					
Phyllodactylus marmoratus		+			+					
PYGOPODIDAE										
Aprasia parapulchella	+	+		+						
Delma impar	+	+		+						
D. inornata	+	+								
Lialis burtonis	+	+			+					
Pygopus lepidopodus	+	+			(+)					
SCINCIDAE										
Carlia tetradactyla	+	+		+						
Ctenotus robustus	+	(+)	+	+						
C. taeniolatus		+		+	+					
C. uber orientalis		(+)								
Egernia cunninghami	+	+	+	+	(+)					
E. saxatilis intermedia					+	+		+		
E. whitii						+			+	
Hemiergis decresiensis	+	+	+	+	+	+		+	+	
H. maccoyi						+	+	+	+	
Leiolopisma coventryi						+	+	+	+	
L. delicata		+	+	+	+	+		+	+	
L. entrecasteauxii (Form A)	+		+					+	+	+

	Grasslands	Savannah woodlands	Lowland swamps	Lowland rivers	Dry sclerophyll forests	Intermediate sclerophyll forests	Wet sclerophyll forests	Upland swamps and watercourses	Sub-alpine woodlands	Alpine communities
SCINCIDAE continued										
L. entrecasteauxii (Form B)						+		+	+	+
L. guichenoti	+	+	+	+	+	+		+	+	
L. metallica							+			
L. mustelina	+	+	+	+		+	+			
L. platynota	+	+	+	+	+					
L. trilineata	+				+			+	+	+
Lerista bougainvillii					+					
Menetia greyii	+	+		+	+					
Morethia boulengeri	+	+	+	+	+					
Pseudemoia spenceri						+			+	
Sphenomorphus kosciuskoi								+		+
S. tympanum (W.T.F.)				+	+	+	+	+		
S. tympanum (C.T.F.)					+	+	+	+	+	+
Tiliqua casuarinae									+	+
T. nigrolutea						+			+	+
T. scincoides	+	+		(+)	+					
Trachydosaurus rugosus	+	+			(+)					
VARANIDAE										
Varanus varius		+		+	+	(+)				
ELAPIDAE										
Acanthophis antarcticus				+	+					
Austrelaps superba						+		+	+	
Cryptophis nigrescens				+	+		+			
Drysdalia coronoides						+	+	+	+	+
Furina diadema		+		+	+					
Notechis scutatus	+	+	+	+	+			+		
Pseudechis porphyriacus	+	+	+	+	+	+				
Pseudonaja textilis	+	+	+	+	+					
Unechis flagellum	+	+	+							
U. gouldii	+	+		+	+					
Vermicella annulata	+	+		+	(+)					
TYPHLOPIDAE										
Typhlina nigrescens	+	+	+	+	+		+			

2
Reptiles at Large

Before we consider the reptile fauna of the Australian Southern Highlands region in detail, we must survey the group in more general terms.

Reptiles have undergone extensive adaptive radiation including arboreal, terrestrial, burrowing and aquatic forms. The present world fauna includes four reptilian orders of which three are represented in Australia. Of these, Chelonia (tortoises and turtles), the Squamata (lizards and snakes) and Crocodilia (crocodiles and alligators), only the first two have representative species in the area treated in this book.

Chelonia

The Chelonia are an ancient group in terms of evolutionary history, and the species retain many of their primitive characteristics.

Typically, the body of chelonians has undergone a shortening and broadening with the development of a bony box into which the head, tail and limbs can be withdrawn. The shell, which is certainly the chelonian's most obvious character, is composed of a dorsal *carapace* and a ventral *plastron*. Each of these is formed of an inner layer of bony plates which are invested to the outside with plates of horny material (*tortoise-shell*). These horny plates are equivalent to the scales of other reptiles.

In all the present day chelonians there is a complete absence of teeth. The margins of the jaws are transformed into sharp ridges which are covered with a horny beak.

The group as a whole contains a wide variety of species found throughout most of the warmer parts of the world. The

33

diets range from carnivorous, through omnivorous to vegetarian. There are aquatic and terrestrial species, and thus a
wide variety of different habitats are encountered.

The characteristics of the Eastern Long-necked Tortoise
Chelodina longicollis are considered in more detail on page 78,
and this species is the sole representative of the group in the
Southern Highlands.

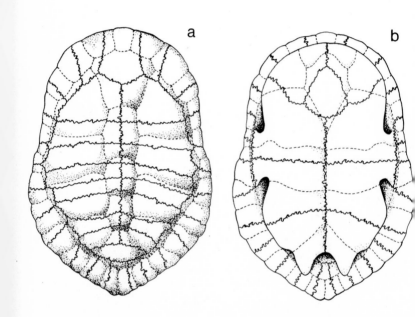

Figure 2
The shell of the Eastern Long-necked Tortoise *Chelodina longicollis*.
(a) dorsal aspect (b) ventral aspect
The major structure, which consists of bony plates, is outlined by a continuous line showing the divisions (sutures) between the plates. On the
dorsal surface, the carapace, there is a paired row of 'rib' plates and a
a pair of marginal rows which encircle the whole. These bones are not
visible in the living animal as they are overlaid by horny scutes (tor-
toiseshell), the outlines of which are shown by broken lines. Note that the
two types of covering are alternate so as to increase structural strength.

Squamata

The Squamata is divided into two sub-orders: Lacertilia (lizards) and Ophidia (snakes). The lizards and snakes may be considered to be the most successful of modern reptiles and number between them some 5000 species. The Lacertilia is regarded as being the more ancient group, appearing in the fossil record in the Triassic epoch, about 170 million years ago, whereas the earliest snake did not appear until the Upper Cretaceous, some 80 million years ago. Even so the latter sub-order did not become abundant until the Oligocene period about 40 million years later.

Although the two sub-orders have many distinctive features, it is believed that the snakes were originally derived from some lacertilian ancestor which lost its legs (amongst other changes) during the course of evolution. The tendency toward limb reduction is widely exhibited in the Lacertilia today and seems to have occurred independently in members of about half the living families. The progress towards limblessness is apparently an adaptive feature in relation to a burrowing habit.

In all snakes and some lizards, the eyelids have become modified to form an immovable transparent spectacle covering the cornea with the loss of the nictitating membrane. This character, which is foreshadowed in some groups of lizards by the development of a window (*palpebral disc*) in the still-movable lower eyelid, may be a pre-adaptation to a burrowing habit. However, the full adaptive significance is not clear as this trait is exhibited by many arboreal and surface-dwelling species. In addition, the eye of snakes appears to have evolved as an entirely new structure and may represent a redevelopment from an eyeless or near eyeless condition. This feature adds considerably to the theory that the snakes were derived from an ancestral burrowing form, particularly as some primitive snakes are both burrowers and blind (viz. *Typhlina*).

The sense of smell is well developed in many species and both lizards and snakes possess a peculiar sensory device known as *Jacobsen's organ*. The organ is a paired structure and consists of two odour-sensitive pits located at the front end of the roof of the mouth. Scent particles are conveyed to these pits from the air or solid objects by way of the tips of the forked tongue which is found in all snakes and in some lizards. The almost incessant flickering of the tongue of snakes and of some lizards, notably the varanids (goannas), and the occasional dabbing at novel objects with the tongue by many other lizard species is the outward manifestation of odour testing. Just as a dog sniffs at a trail, so can a snake find its prey by means of its tongue and Jacobsen's organ. The tongue is seen to be most active when an animal is alarmed, is identifying its prey, or during examination of a mate at courtship.

Control of body temperature can be a problem for so-called 'cold-blooded' animals. The term 'cold-blooded' can be misleading, as a snake or lizard basking in the sun on a rock can have a blood temperature approaching that of the rock itself, which may be as high as 40°C.

Both lizards and snakes use a wide variety of methods for maintaining their body temperatures within workable limits. The most obvious temperature-regulating mechanisms are behavioural. Sunbasking to gain heat or shade-seeking to lose excess heat are the most common methods, but a number of refinements are used in addition. In gaining heat, an individual may expand, flatten and angle its body toward the sun or may closely adpress it to a warm surface (such as a re-radiating rock or road surface after sunset). The expansion and flattening process may expose darker areas of bodily coloration which aids in absorbing heat, or the animal may undergo rapid darkening of its general coloration. The latter is most commonly seen in the larger agamids (dragons), particularly of the genus *Amphibolurus* which locally includes the common Bearded Dragon. Heat loss may also be accomplished by a variety of means; simply for an animal to avoid direct radiation may not be sufficient. Thus, body postures which include angling to present the smallest area to the sun, colour changes involving lightening of hue to aid in heat reflection, gaping and ventilation of the mouth cavity resulting in evaporative heat loss, and pneumatic inflation of the body thus exposing thinner areas of skin between the scales are all employed, either singly or in various combinations.

High body temperatures can be maintained for long periods within a few degrees of a preferred level, which is generally the optimum for the continuance of normal behavioural activity. Such stability within fixed limits is only possible as long as the ambient temperature does not fall below the lower limit and the opportunity for sunbasking is available.

Broadly speaking, reptiles may be separated into two groups on the basis of the method used for maintenance of their body temperature during normal behavioural activity.

(i) **Heliotherms** Basking reptiles which use solar radiation to raise body temperature, enabling them to exploit areas of low ambient temperature.

(ii) **Thigmotherms** Non-basking reptiles which gain heat from their immediate surroundings, enabling them to inhabit shaded areas but generally restricting them to warmer environments.

In an extreme winter climate, such as that experienced in the Southern Highlands, the mean daily temperature is not high enough to ensure heat gain to a level required for activity. Hence, all the local species will enter a period of torpor for a length of time which is directly related to the length of the colder weather experienced by each in its range. This period of torpor is usually entered before the onset of the earliest cold winter weather. The animal has by this time found a natural crevice or cavity beneath a stone or log or has excavated a deep burrow. Some species will occupy these retreats all winter and only emerge with the start of the warmer weather the following spring. The heliotherms, being able to exploit solar radiation, sometimes break their winter torpor during a mild spell and may be seen basking near their burrows. Generally, the breaking of winter torpor occurs when the 'hibernation' site is locally warmed by the sun. Each species thus tends to have a preferred emergence temperature when ambient temperatures exceed the lower limit of the behavioural range.

The skin of reptiles is characteristically dry and contains either a few glands or none. The upper layer of the epidermis produces the horny scales which are so characteristic of lizards and snakes. Periodically the outer layers of the skin including the upper keratinised layer of the scales is shed, either in flakes as is common for most lizards, or as a single *slough* typical of the snakes.

The process of sloughing is intimately associated with the growth of the animal. The skin is only of limited elasticity, so growth must be accompanied by the periodic occurrence of the slough to allow for a fresh and slightly larger area of skin surface. The growth rate is sporadic and largely depends upon the availability of food, the age of the animal and its general health and condition. Growth rates generally appear to be greatest in young animals, slowing gradually towards maturity, although the annual increments may even then appear to be large. For instance, a mature Green Python in captivity grew from 152 centimetres to 179 centimetres in two years representing an annual increment of over 13 centimetres. Statistics of this kind, however, tend to be biased upwards when based on captive specimens as such animals are not only usually kept well fed but are cushioned from the rigours of the natural environment.

This leads to the consideration of the longevity of reptiles. Unfortunately little is reliably documented, and even then the information is subject to qualification. Inevitably, recorded life-spans are for individuals which have been maintained in captivity under artificial conditions. They have had regular and good-quality diets, even 'climatic' conditions, treatment for ailments and protection from parasites and predators. It is doubtful whether the longevity records for these animals bear any similarity to the natural life expectancies of their wild counterparts. However, we can say that amongst the reptiles are some of the longest living of all terrestrial vertebrates, and that the larger species normally tend to be the longer lived. For instance, the Galapagos Tortoise is recorded as surviving for well over 100 years.

As the two sub-orders are quite distinctive in most of their attributes, they will be considered separately.

Lacertilia

Most lizard species are quadripedal although, as was mentioned on page 35, there is some tendency towards limb reduction. This tendency may be manifested as a reduction in the size of limbs, loss of toes, or complete loss of the forelimbs and hindlimbs. In Australia, the tendency to limb reduction is found most generally in certain skink genera and culminates in the almost entirely limbless (*apodous*) condition of the family Pygopodidae. The latter family is represented locally by five species. In these, the hindlimbs are represented by flap-like vestiges which are held into lateral grooves and thus do not project beyond the general body contour.

A peculiar feature of many lizards is their ability to drop their tails when alarmed. This is accomplished by a specialised adaptation of some of the caudal vertebrae across each of which there is a predetermined plane of weakness. Pressure on the tail by a would-be predator or even voluntary contraction of the tail muscles can cause the tail to break off. This trait is known as tail *autotomy*. Autotomy appears to do little harm to the lizard which grows a new tail, although the regrown portion only possesses a simple cartilaginous support and is incapable of further autotomy. Any additional voluntary breaks can then only occur on the remaining (*proximal*) portion of the original tail. The scale patterning and coloration of the regrown portion is usually different from that of the original tail and it is usually shorter and less well proportioned.

The ability to autotomise the tail confers considerable benefit on the animal when it is attacked; a predator is usually distracted by the vigorously-squirming, dropped appendage and the former owner can make good its escape. The significant survival value of this characteristic goes without saying;

many species which possess this attribute will, in the presence of a predator, draw attention to the tail by vigorously moving it about. In many species the tail is also brightly coloured or obviously marked in distinction to the remainder of the body (viz. *Ctenotus taeniolatus*, p. 133).

The four local pygopodids, both geckoes and all the skink species, with the exception of the Bluetongues and the Stump-tailed Skink, have the ability to shed the tail voluntarily.

As the ordinal name indicates, lizards are invested with a skin of horny scales which are frequently imbricated, that is, they are disposed like the overlapping tiles on a roof. The scales may be smooth or keeled (*carinated*). As was mentioned on page 38, sloughing of the skin by lizards tends to be by small fragments, but this is not so for the geckoes and the pygopodids which usually cast the skin complete.

The ear opening is usually external with the tympanic membrane exposed and immediately discernible, although in some specialised forms this is subject to modification.

The lower jaw of lizards has a fixed articulation, in contrast to that of snakes (see page 48), so large food must be dismembered by biting, clawing or worrying. Many lizards are omnivorous; few are either wholly vegetarian or carnivorous. Most lizards display a limited flexibility of the skull at the frontoparietal suture which permits the swallowing of large prey.

The diet usually consists of insects or other small arthropods, although larger species may sometimes consume vertebrate prey. The tongue is frequently used as a means of catching small food and is, in some forms, specialised to accomplish this in combination with a viscous, sticky saliva.

The local geckoes, pygopodids, most of the snakes and the majority of the agamids subsist predominantly on arthropod (insects, spiders, etc.) prey. The larger agamids, such as the Gippsland Water Dragon *Physignathus l. howittii* and Bearded Dragon *Amphibolurus barbatus*, will readily feed on vertebrates including young of their own species and other small reptiles. The only local varanid *Varanus varius*, like most other species within the family, is a frequent carrion feeder and may often

40

be seen tearing at animal corpses on the roads. *Varanus varius* will also take an assortment of living vertebrate prey including young birds and eggs; young rabbits and mice also form a significant component of the prey, which makes the conservation of this species an important consideration.

In spite of the readiness of most species to feed on animate prey, there is considerable supplementation of the diet with green vegetation. Some of the local agamids and larger skinks have a predilection for yellow flowers and these they will consume with great avidity. An assortment of fruits, particularly blackberries, are also taken and individuals in captivity often display a liking for banana, apple and other soft fruits.

Fertilisation in reptiles is internal. In all lizards, the opening of the vent is transverse and the male copulatory organ is a paired structure, each lobe of which is known as a *hemipenis*. The hemipenes normally lie within the tail behind the vent, where they may cause a slight swelling at the base of the tail. In some species, the presence or absence of this thickening can be used to diagnose sex (this is most noticeable in the geckoes) but, as it is a variable feature between species and may often depend on the state of obesity of the individual, its use as a sexual diagnostic character is of doubtful value. The hemipenes are composed of erectile tissue and during copulation one or other is everted through the vent on either side of the base of the tail. In a freshly dead animal, the hemipenes can be everted by applying pressure behind the vent. An attempt at eversion in this manner with a live individual could result in rupture of the organs and is thus inadvisable. In some species the hemipenes are ornamented, to a greater or lesser extent, with spiny protuberances. The relative elaboration of these ornamentations are unique to each species and could, in the wild, help to prevent hybridisation. Perhaps a more likely explanation is that they may help to maintain the hemipenes in position during sperm transfer. During mating the male usually uses only one hemipenis. After mating the sperm are stored in the reproductive tract of the female often for considerable lengths of time until ovulation occurs and fertilisation follows.

Courtship behaviour is usually elaborate. In the agamids, rapid colour changes are employed by the male, but in all groups there are specialised behaviour patterns which usually involve postural changes. These postures may display areas of body pattern and coloration which are generally accentuated during the breeding season. The gular fold in many species is often inflated or raised and in a large number of skink species the throat of the male becomes strongly and distinctively coloured in season (pages 168, 179). The Bearded Dragon *Amphibolurus barbatus* uses its elaborate gular pouch, not only in the familiar threat display, but also in courtship sequences. Head bobbing, arm waving and lashing of the tail are also commonly used. Other postural changes include expansion, flattening and angling of the body to present the largest area to the prospective mate. It is interesting to note that very similar behavioural sequences are employed during threat or aggression. In fact the initial approaches of the male are characteristically those of a threat display (see page 45), and the continuation or development of a courtship sequence usually depends on the first and subsequent responses of the female.

Most lizards are oviparous, that is they lay typical shelled eggs, and, while some may temporarily guard the nest burrow, there is no parental care of the young.

Among the local families, all the dragons, both geckoes, the pygopodids and most of the skinks are oviparous. The eggs are usually deposited within a subterranean chamber connected to the surface by a narrow tunnel which is sealed after egg laying. The egg chamber is commonly, though not necessarily, beneath a stone which aids in providing an efficient heat reservoir for incubation. The number of eggs for each species varies between two and 20 or more.

The egg shells of oviparous lizards range in their degree of thickness from a thin, parchment-like membrane through to the tough, calcified shell of some geckoes. Non-calcified eggs are readily subject to desiccation, so their deposition in a burrow protects them not only from extremes of temperature and the ravages of predators, but also from the possibility of drying out.

The stage of development at which eggs are laid varies from species to species. They may be laid soon after fertilisation has occurred, as in the case of those species inhabiting more temperate regions, or at various later stages up to the time of hatching. The extreme of the latter condition is similar to the next mode of reproduction.

In a few genera ovoviviparity is common, that is, the eggs are retained within the female's body until hatching. This mode of reproduction is displayed most commonly by those groups that inhabit colder climates, and may be an adaptation for ensuring adequate incubation for the eggs as the female seeks to regulate her own body temperature.

In some species, there is the further specialisation of the development of a placental attachment via an umbilical cord which allows close association between maternal and foetal circulatory systems, with the passage of nutrients to and wastes from the developing embryo. This evolutionary development, which is normally associated with the mammals, might appear to be a remarkable attribute for the reptiles. It is evident that the condition has evolved independently on a number of occasions in different groups throughout the evolution of the reptiles. It is evident locally in the Bluetongues and Stump-tailed Skink.

Threat and aggression

As mentioned above, elements of aggressive behaviour are often incorporated into the courtship displays. However, to consider this behaviour, threat and aggression must be divided into two components: *intra*specific, where the animal responds to others of its own species; and *inter*specific, where the animal must usually deal with a would-be predator (including man).

Intraspecific behaviour patterns are highly ritualised. For example, male Bearded Dragons are markedly territorial during the breeding season, and at other times, in high density situations, they also establish dominance hierarchies. When an encroachment is made by an outsider into the territory of a male, the owner of the territory initiates the aggression ritual by lowering his gular pouch so that it is vertical beneath his jaw. This attitude is very different from when he deals with a predator, as shown below. At the same time he undergoes a rapid change of colour involving blackening of the gular pouch and of the last one-third to half of the tail. He then advances towards his adversary who may either flee or assume a similar posture while continuing his advance.

Throughout this initial approach, both individuals may pause repeatedly and beat the ground with one or both forefeet. The head may also be bobbed[1] vigorously with a fast down movement and a slower upward one.[2] When the two animals come into close proximity, they circle one another slowly, head-to-tail, and at the same time flatten and angle their bodies, thereby presenting the greatest area to one another.

The next phase entails vigorous sideways lashing of the tail in attempts to beat the head of the opponent, but simultaneously trying to grasp the base of the opponent's tail in the mouth. If one or other manages to bite his opponent's tail, he tries to turn him over with a quick sideways jerk of his body. However, this rarely succeeds and the two break and recommence their circling. This sequence usually continues until one individual flees, quite often to be pursued by the victor. The tail biting usually only results in temporary surface wounding and the adversaries are never permanently injured, apart from very minor scarring.

It is interesting to see elements of the courtship sequence in these disputes. Sometimes the vanquished male will exhibit

[1] Head bobbing is one of the most common of all display techniques used by a wide range of reptile species from different families throughout the world.

[2] Variations to this sequence are: slow downward movement followed by rapid upward one — *Physignathus l. howittii* (Gippsland Water Dragon); both movements rapid and repeated in staccato bursts — *Amphibolurus muricatus* (Jacky Lizard).

the typical appeasement display of the sexually submissive female by slow circling of the head accompanied by slow, but jerky, circling of one or other front leg. Evidently, in both instances this behaviour serves to terminate aggressive behaviour in the male, for in the case of courtship, the male mounts the female at this juncture and will likewise attempt to mount the vanquished male.

It is also interesting to see very similar rivalry displays between females, or between a female (particularly if she is gravid) and a younger male.

The interspecific display, again using the Bearded Dragon as the example, presents an entirely different pattern of behaviour.

Upon a close approach by a predator, an individual of either sex will face its adversary and expand the gular pouch laterally, while simultaneously gaping and thus exposing the bright yellow interior lining to the mouth. This part of the display is frequently accompanied by loud hissing which is effected by vigorous expulsion of air from the lungs. Should this temporarily hold the would-be predator at a distance, the lizard takes advantage of this pause to make repeated gulps of air, using the gular pouch as a bellows, and thereby re-inflates its body. This inflation serves several functions: firstly, the rubbery air-filled body is difficult to grasp; secondly, the hard, though usually mobile, spiny scales on the sides of the body become erect; thirdly, the lizard simply looks larger and more daunting; and fourthly, should all this still fail to deter the predator, the air provides a large reservoir for renewed and vigorous hissing.

As a last resort, the lizard may simply try to flee or may even leap towards the predator with mouth agape. Although usually disinclined to bite, the Bearded Dragon, like most other species, will ultimately resort to this expedient if provoked.

Perhaps it should be emphasised here that there are no venomous Australian lizards. However, inexpert handlers of lizards must take extra care with the larger species as not only are many capable of inflicting a hard and often painful bite, but if they should break the skin there is the strong possibility of infection entering the unclean wound. The latter should be remembered with particular regard to those species which often feed on carrion.

It is noteworthy to compare the use of the forms of gular pouch expansion alluded to above. In both intra- and interspecific displays, the larger area of the expanded 'beard' is exposed to the adversary. In the case of intraspecific rivalry, the head is kept sideways on and the pouch simply lowered below the jaw; whereas, in the alternative case, the lizard faces the attacker and expands the pouch laterally.

Although the Bearded Dragon has been chosen as the example here, close study reveals similar behavioural characteristics in other species with head bobbing and foot stamping, gaping and hissing being common components.

Figure 3
Rivalry between two male Bearded Dragons *Amphibolurus barbatus*. The two antagonists circle one another, each attempting to lash the other over the head with its tail or trying to grasp the base of the opponent's tail in the mouth. Note the posture of the blackened beard in the nearer individual and the distribution of dark pigment in the tail.

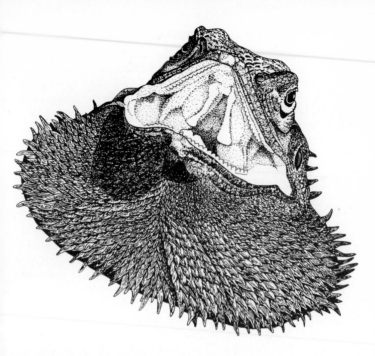

Figure 4
The threat display of the Bearded Dragon *Amphibolurus barbatus*. The mouth is held agape so displaying the bright yellow interior which contrasts with the blackened, expanded beard. The beard is elevated by means of the paired hyoid bones of the jaw, and in this position presents its widest expansion to the intruder.

Ophidia

The second sub-order of the Squamata is Ophidia, the snakes. Perhaps the most prominent characteristics of the snakes are their elongate, limbless and sinuous form, their unblinking gaze, and their restless, flickering tongue. In addition, most people think of snakes as being generally venomous and thus treat them with considerable circumspection or acute distaste. Probably no other group of animals, with the possible exception of the spiders, elicits so much horror in so many. In spite of the fact there there are many quite harmless species in the world fauna, Australians have every right to be wary of snakes as most of their larger genera belong to the highly venomous family, the Elapidae.

In the Southern Highlands, there are four large, common species of this group worthy of special note. They are the Brown Snake *Pseudonaja textilis textilis*, the Mainland Tiger Snake *Notechis scutatus scutatus*, the Copperhead *Austrelaps superba*, and the Red-bellied Black Snake *Pseudechis porphyriacus*. All of these species can be dangerous to man.

Apart from the most obvious characteristics of snakes which were referred to above, the group exhibits some unusual and interesting specialisations. First and foremost must be the remarkable adaptations of the skull and jaws which enable these exclusively predatory animals to distend the mouth to an enormous degree and to swallow prey which is considerably larger in diameter than the snake itself. Progressive modification of the skull and jaws throughout their evolutionary history has resulted in not only the loosening of the ligaments of the lower jaw and its articulation with the skull (as in some lizards), but the breaking up of the elements of the skull itself into flexibly-jointed pieces which articulate and move to permit this phenomenal swallowing capacity. Complementary to this development has been the progressive modification and specialisation of the teeth, accompanied in the more advanced groups by the addition and refinement of the venom apparatus.

In the more primitive families, Boidae and Typhlopidae (only the latter is represented locally by the Blind Snake *Typhlina nigrescens*), there are still vestiges of the ancestral pelvic girdle; in the case of the boids (pythons), the bones of the hind limb are represented by two minute, vestigial, external claws. In these, as in all other families, all traces of the front limbs and their pectoral girdle have been lost.

In association with the evolution of the elongate form of snakes, some internal organs have undergone adaptive modification to the extent that only one lung is developed in most species.

When introducing the lizards, the unusual and novel structure of the snake eye was mentioned (page 35). Many of the original components of the reptilian eye have been lost, but they have been replaced by similar structures derived from new sources. In addition, there is neither eyelid nor nictitating membrane and the outer surface is covered by a transparent spectacle which is part of the general skin surface. As a consequence, the spectacle is lost along with the rest of the skin each time the snake sloughs. As the time of the slough approaches the snake seeks a place of retirement as the whole of the eye becomes opaque during the separation of the old spectacle from the new. At this time, and until sloughing has been accomplished, the snake is virtually blind.

The typhlopids, which are habitual burrowers, have degenerate eyes, the positions of which are marked by dark spots located on the ocular scales of the head (Figure 7).

The ophidian ear has also undergone considerable modification and reduction. There are no eardrums, external ear openings or Eustachian tubes. It is unlikely that snakes are able to hear airborne sounds, although they are highly sensitive to ground vibrations which are picked up and transmitted by the bones of the jaw.

The highly adaptive and useful lacertilian character of tail autotomy is not possessed by snakes, although accidental breakage can occur when an animal is restrained or suspended by its tail. There is no regeneration of the lost portion.

There are four methods of progression normally employed by snakes. Most common is the typical sinusoidal, side-to-side motion. This most efficient on rough surfaces or in grass and other vegetation. On loose and sandy surfaces, this method is replaced by a remarkable sideways movement in which the snake progresses on a sort of rolling spiral with the body in contact with the surface at only two points. This second form of locomotion is referred to as *sidewinding*, and is rarely seen in any of the Australian species. It is most highly developed in some desert-dwelling species of North American rattlesnakes.

The third method is the concertina principle which is employed only on smooth surfaces, where there are few points to which the snake can anchor with its ventral scales. The forepart of the body is pushed forwards in a sideways loop until an anchor point is secured. The hind end is then pulled up after it.

The last type of movement is on the principle of caterpillar progression and is referred to as *rectilinear creeping*. The body is held in a reasonably straight line and the individual ventral scales are moved forward in sequence, each becoming anchored and allowing the animal to pull itself along. The general movement of the animal using this mode of progression is almost imperceptible and the overall impression is of gentle gliding. Many of the larger snakes, principally the pythons, use this type of locomotion, when it may be most commonly used by an animal stalking prey. Fast movement is impossible by this method and, for the local species at least, only the first method is used for fast turns of speed.

The disposition of the body scales in snakes differs from the lizard condition. The dorsal complement are usually small, smooth and slightly imbricated, whereas the ventral surface is invested with a single row of broad overlapping plates. The rear margins of these belly scales are sharp and may be used to anchor the snake as it moves.

It has been stated above that the skin is shed whole at each sloughing. During the slough, the skin first splits around the margins of the jaws, then the snake draws itself through the skin which pulls backwards, turning inside-out as it progresses. The animal often uses vegetation or stones to anchor the slough as it draws itself out. The actual process of sloughing may take anything from a few minutes to several days.

Some groups of snakes have developed a remarkable sensory device for finding warm-blooded prey. The organs concerned are infra-red receptors and are located on either side of the head. In the pit-vipers and rattlesnakes, these are simple pits; in the pythons, they are represented by a complicated-looking structure on either side of the lower jaw. The sensitivity of these facial and labial heat sensors is such that they are capable of discriminating between very small temperature differences within the immediate surroundings of the snake.

Snakes feed on live food only and prey may consist of anything from insects to large vertebrates in a maximum utilisation of all types of animate food resources, although a few species are specialists in their dietary requirements. The prey is usually swallowed whole and, although there may be many disadvantages in consuming a quantity of often bulky prey at the one time, it does allow the snake to go for very long periods between meals. Some snakes can swallow struggling prey with apparent ease, but others have evolved various specialised ways to quieten their victims first. This may be accomplished either by suffocation as by the pythons, or by poisoning as is done by the majority of the Australian snakes.

The pythons, the suffocators, will strike their prey and maintain their grip while wrapping their body coils around the writhing victim. This accomplished, they then squeeze the victim until it can no longer breathe; there is generally no crushing. The prey is then often released while the snake examines it closely with its tongue. Ultimately the head of the prey is found and swallowing begins.

The poisoners, on the other hand, rely upon the effectiveness of their venom to quieten prey. The nature of snake venoms is discussed in Chapter Five on page 247. Note that the venom itself, which is really a potent saliva, is probably primarily a prey-quietening substance and has been evolved only secondarily as a self-defence attribute.

The venom glands themselves are developed from modified salivary glands but, in addition to the various toxins produced, the glands still supply digestive enzymes as in the non-venomous ancestral forms. It is curious to think that a venomous snake, once having struck at its prey, can not only sit back and wait for its victim to die, but also expect a convenient, partly pre-digested meal.

Among the world venomous snakes, there have been at least two evolutionary trends in the refinement of the venom technique. In the vipers and rattlesnakes, not represented in Australia, there has been considerable sophistication in the development of erectile, tubular fangs for delivering the venom. The venom itself in these groups, although highly potent, has not reached the degree of chemical complexity of that of the Elapidae, which includes the Asian cobras, kraits, African ringhals and mambas and most of the Australian venomous snakes. In these snakes, the fangs are short and fixed. The progressive elaboration of venoms represents the second trend.

A classification of snakes based on the structure of the teeth figures prominently in scientific writings. The relevant terms are defined as follows:

Aglypha   Non-venomous snakes. Although some of the teeth may be fang-sized, they are all solid and not grooved and canalicular.

Opisthoglypha   Usually mildly-poisonous snakes with one or more pairs of simply-grooved venom fangs on the top jaw, *preceded* by a number of solid teeth.

Proteroglypha   Venomous snakes which have one or more pairs of rigidly-erect venom fangs on the maxillary bone, *preceded* by a number of much smaller solid teeth. The fangs are grooved (but functionally hollow) to allow the injection of the venom. This group includes two families, the Elapidae and Hydrophiidae (sea snakes).

Solenoglypha Venomous viperine snakes, which include the families Viperidae (true vipers) and Crotalidae (pit-vipers). The venom fangs are much longer than in the preceding group and are attached to movable bones of the skull which enable them to be retained against the roof of the mouth when the mouth is closed. When the mouth is opened for the snake to strike, the fangs are erected by a system of levers through the articulations of the movable bones. By a progressive infolding of the margins of the grooved fangs in ancestral forms, each fang has become canalicular and acts physically like a hypodermic needle.

In terms of numbers of species of venomous snakes, Australia and Africa tie at third place with between 70 and 80 species each, led by the Americas with nearly 100 species, and Asia with a still greater number. In the Southern Highlands, the snake fauna includes 12 species of which all but one are venomous.

The use of venom as a defence mechanism is not the only device that a snake has for escaping injury or being preyed upon by other animals. The most common protective device is that of fleeing when disturbed, and the majority of snakes will do just that when alarmed. Under such circumstances snakes exhibit surprising turns of speed. It is probably this factor alone, in a country as richly endowed with common, large, venomous species as Australia, which accounts for the low incidence of snake-bite in man — but more of that in Chapter Five on page 247.

The second defence mechanism is camouflage. Many snakes are remarkably well concealed in nature although, when seen in captivity, they appear to be highly patterned and brightly coloured. Many will simply escape detection by remaining immobile.

A third defence is for the snake to form a tight ball with its head tucked well inside the coils of the body. This is very commonly used by many of the non-venomous species.

A fourth type of defence, often used by non-venomous species, is sham striking. This relies upon the fact that venomous snakes will strike at an attacker, and, in mimicking this, the non-venomous snake derives real advantage from this ploy, although it does not have the necessary apparatus to back up its display. At the same time it may be able to inflict wounds with sharp teeth and these (as for the larger lizards) can become infected.

In addition to these four, there are a number of specific defence mechanisms which will be discussed later for individual species.

The snake will generally use venom as a defence only as a last resort. Upon reflection it might seem unwise for an animal to squander too freely its only effective means of prey-catching. Not only is the amount of venom often limited, but the teeth are delicate and are liable to be damaged by striking at large adversaries. For these reasons, most of the venomous species resort to elaborate warning displays which include sham striking. The classic example of this behaviour is shown by the King Cobra, but the same type of display is commonly seen in its Australian familial relative the Tiger Snake. The display characteristically includes flattening of the neck, rearing up of the front portion of the body, swaying and hissing.

Fertilisation is internal for snakes as it is in the lizards. The vent is transverse and, in the male, the hemipenes lie in the base of the tail behind the vent. There is no obvious bulge caused by the presence of the hemipenes. There is considerable variation in the spinose elaboration of the hemipenes and these, as in the lizards, probably serve to anchor the hemipenes during copulation. Some female snakes are able to store viable sperm in the reproductive tract and thus delay fertilisation for long periods after copulation.

In spite of frequent references in the literature to elaborate courtship displays by snakes, few observations have been scientifically documented. Copulating snakes are often found intertwined, but there is little known about the preliminary behaviour.

Most snakes are oviparous, but few show any parental care of the eggs or brood. The eggs are laid in some protected spot, usually in a convenient crevice in the base of a tree, amongst or beneath rocks, or in a burrow. In some species, the female remains coiled about the egg clutch, probably not for incubation but simply for protection of the eggs against predators.

A few species are ovoviviparous and, like those lizards showing this characteristic, have a distinct preference for severe climatic habitats. Thus it is understandable that most species common to the Southern Highlands are ovoviviparous.

Young venomous snakes, which can sometimes be more beligerent than the adults, begin life with a fully operative venom apparatus. However, it functions on such a small scale that, although they may be fully capable of subduing small prey, they are generally unable to inflict serious damage on larger predators.

3
Classification, Nomenclature and Diagnostics

Introduction

Before any study of the biogeography, ecology, physiology or behaviour of a particular kind of animal or group of animals is embarked upon, it is necessary to give a clear descriptive definition of that animal and of others which are closely associated with it.

While it is clear that many animals differ from each other, it is just as obvious that many also closely resemble each other.

In this way, they can be classified into groups and subgroups according to their structural affinities or differences. Each of these groupings is assigned a name, the *taxon*, and the process of description involves a formalised system of nomenclature.

Classification may be considered to consist of three elements: *hierarchy, distinction* and *grouping*.

Where two kinds of animal differ from one another in some well defined but perhaps relatively minor structural character or characters, they may be regarded as being distinct *species*. The status of species for both kinds is confirmed if there is found to be a barrier to interbreeding between them. This *reproductive isolation* may be achieved by a variety of means which need not be directly related to the morphological differences. Species are grouped into *genera*; a *genus* is an assemblage of species showing evidence of close relationship in common characters. When more is known of a complete assemblage of species, certain of these species may be found to have characters which show no close affinity with any other species in the group. The taxonomist may therefore find it necessary to propose a new genus to contain the one distinct species; such genera are termed *monotypic*.

A genus in turn is classed into a wider category, the *family*, whose component genera all display some important characters in common. All families which possess major linking features are grouped together in an *order*. Then orders collectively form a *class* and classes are grouped on the basis of common fundamental characters into a *phylum*. Phyla collectively form the *kingdom* of animals.

In some large and complex groups of animals, certain intermediate taxa are adopted. The most taxonomically reliable of these sub-groupings are: the *sub-family*, a section of a family containing a group of genera; the *super-family*, a group of families below the status of an order; the *sub-order, super-order* and *sub-class*. Perhaps the least reliable of these sub-groupings in general is that of the *sub-species*, where relatively minor distinguishing characters (e.g. colour; see page 60) may not express significant phylogenetic distinction between populations.

Set out below is an example of the hierarchical system of nomenclature for the Stone Gecko, *Diplodactylus vittatus* Gray.

KINGDOM	Animalia
PHYLUM	Chordata
SUB-PHYLUM	Vertebrata
CLASS	Reptilia
SUB-CLASS	Diapsida
SUPER-ORDER	Lepidosauria
ORDER	Squamata
SUB-ORDER	Lacertilia
INFRA-ORDER	Gekkota
FAMILY	Gekkonidae
SUB-FAMILY	Diplodactylinae
GENUS	*Diplodactylus*
SPECIES	*vittatus*

The primary purpose of taxonomy is to define the distinctions between species and their groupings or affinities. It may deal with concepts only, as an order, a family or a genus may merely reflect the taxonomists' views on the grouping of various species to show their relationships. In some cases it may be desirable to construct this classification using the phylogenetic relationships of animals as a means of portraying their evolutionary history.

Whatever the purpose of classification may be, the first step must be to attach a name to each species to use for future reference. The system of nomenclature in universal use is the *binomial* system which dates from the publication of the tenth edition of the *Systema Naturae* of Linnaeus in 1758. The procedure adopted is regulated by international agreement under the *Code of Rules of Nomenclature*. One recommendation of the *Rules of Nomenclature* is that the scientific names of animals must be Latin or latinised words, or considered to be such when not of classical origin. It should be noted that the common endings of familial and sub-familial names are '*-idae*' and '*-inae*' respectively, and that the initial letter of the generic epithet is always a capital while the specific epithet has a lower case initial. When the name of the author is quoted after a specific name it follows in Roman type without punctuation, thus '*Diplodactylus vittatus* Gray'. When a species is transferred to a new genus, other than that in which it was originally described, the name of the author of that species is then given in parenthesis, thus '*Testudo longicollis* Shaw' becomes '*Chelodina longicollis* (Shaw)' following generic revision.

Another aspect of classification is that of the *type*. The type is the specimen from which the published description of the species has been drawn up, and it remains preserved and deposited in a museum collection as the final reference in matters of doubtful identity. The place from which the type specimen was collected is known as the *type locality*.

In field guides, morphological characters will obviously be more useful than the less accessible characteristics such as behaviour, biogeography and reproductive isolating mechanisms. The latter attributes can only be evaluated after exhaustive field and laboratory studies. Because of the reliance upon museum types as the final reference point for species identification, morphological characters are those most usually selected in the construction of identification keys. Even so, a knowledge of the additional characters will elaborate or confirm initial definitions of what constitutes a species, but they cannot easily be preserved as types (if at all).

A further concept in animal classification is the *key*. The key is designed to facilitate the identification of the species to which a particular specimen belongs. The key may be simple or complex. If it is complex, it may use all the characters employed in determining the affinities and distinctions between species. The most common type of key is the *dichotomous* key. In this key, the animals or groups of animals are distinguished from one another by means of a yes/no choice between two sets of characters set out in couplet form. In the lower hierarchical levels of the classification, the range of characters used is often much narrower than those used for determining the higher levels.

Classification of reptiles

The taxonomic characters most commonly used are scalation, dentition, and skeletal structure (in particular the component bones of the skull and their relative juxtapositions). Attempts to use coloration and biogeographic distribution in defining species have led to considerable confusion in the past. An example of the variability of coloration of a cosmopolitan species is that of the pygopodid *Lialis burtonis* which occupies a range of habitats throughout mainland Australia from the tropical to the temperate regions to the arid zones of the interior. In the absence of knowledge about pre-mating isolating mechanisms between populations distinguished on the basis of colour, it would be misleading to assume that these colour variations represent true specific differences.

The use of geographic distribution as a diagnostic of species, and more particularly of sub-species, is also of doubtful value. Careful study usually reveals that what had been regarded as a geographically isolated sub-species is in fact part of the extended range of a single species. Individuals from extremes of habitat type in such a wide range of distribution may show

60

considerable variation in colour, form and size but, in the absence of positive reproductive isolation between these forms, they must all be considered to be part of a morphological range of the one species. However, the phenomenon of speciation is a dynamic process, so populations which represent this *clinal variation* may ultimately evolve into distinct species.

As in any other scientific field, the principles of description demand that there should be a strict system of specialised terminology. A thorough knowledge of these terms is necessary for the successful application of identification keys and to follow species descriptions like those in the latter part of this book. The diagrams (on pages 62 to 67) are provided to acquaint the reader with the diagnostic features of reptiles and their associated nomenclature. The glossary at the end of the book (pages 256 to 268) will clarify the meaning of terms that is not covered in these diagrams.

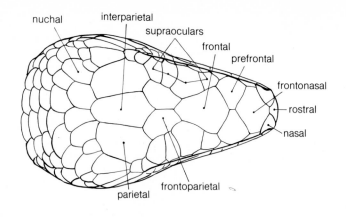

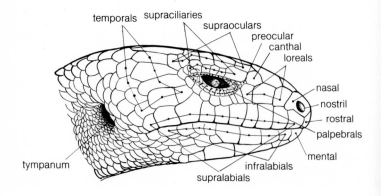

Figure 5
Dorsal (top) and lateral (bottom) aspects of the head of the Common Blue-tongue *Tiliqua s. scincoides* illustrating the arrangement and nomenclature of the cephalic scales.

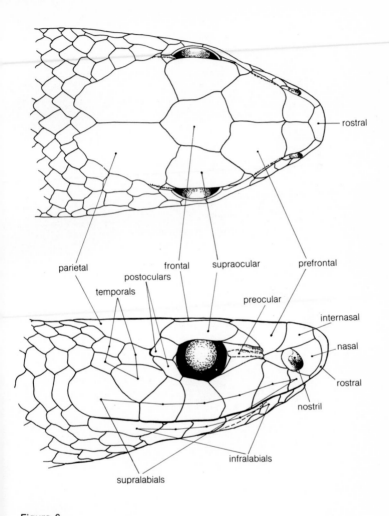

Figure 6
Dorsal (top) and lateral (bottom) aspects of the head of the Black Snake
Pseudechis porphyriacus illustrating the arrangement and nomenclature
of the cephalic scales.

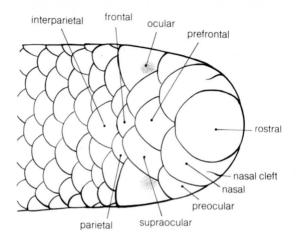

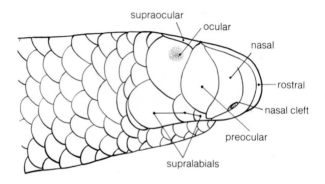

Figure 7
Dorsal (top) and lateral (bottom) aspects of the anterior region of the Blind Snake *Typhlina nigrescens* showing the arrangement and nomenclature of the cephalic scales.

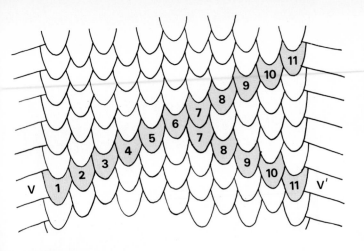

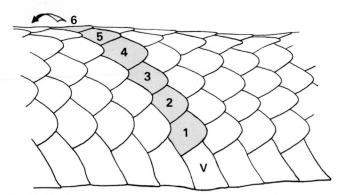

Figure 8

Midbody scale counts on the snake. There are two methods of counting scale row numbers on the midbody region of a snake. Note that the ventrals are not included in the count.

(a) Beginning on a body scale immediately above and adjacent to a ventral scale, count anteriorly along a diagonal row over the dorsal surface of the body until another ventral scale is contacted lying anterior to the first.

(b) As in (a), count anteriorly along a diagonal row to the median scale on the dorsum, then continue to count posteriorly on the opposite side and adjacent to the starting ventral scale (or its paired scale).

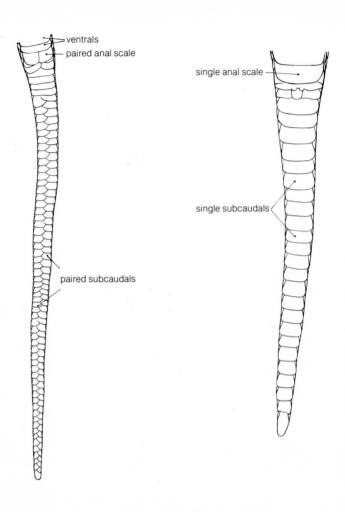

Figure 9
Ventral aspect of the caudal regions of the Common Brown Snake *Pseudonaja t. textilis* (left) and the Black-headed Snake *Unechis gouldii* (right) illustrating the two conditions of the anal and subcaudal scales.

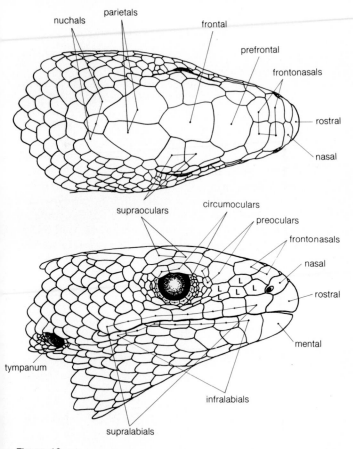

Figure 10

Dorsal (top) and lateral (bottom) aspects of the head of the Common Scaly Foot *Pygopus lepidopodus* illustrating the arrangement and nomenclature[1] of cephalic scales. The group of scales anterior to the eye (lateral aspect) denoted by the letter 'L' are the loreal scales.

[1] Nomenclature after Kinghorn (1926) 'A brief review of the Family Pygopodidae.' *Rec. Aust. Mus.* **15**:40–64.

Notes on collecting

Reptiles in captivity

The keeping of live animals can be a rewarding experience in many ways, and is likely to be highly educational. However, a note of caution must be sounded to the would-be collector as existing legislation in New South Wales[1] and proposed legislation in Victoria[2] and the Australian Capital Territory[3] restricts or disallows the capture and retention of most reptile species. In many ways this legislation is commendable in that it seeks to protect rare and endangered species, but it could severely limit the opportunities for the inquisitive child to gain first-hand experience of a wild animal. On the other hand, many species of reptile do not readily adapt to captivity and may sicken and die even when the correct diet is available in quantity and in the most experienced hands. Within the law, it is possible to keep limited numbers of certain species, and the serious student will usually be granted a collector's permit.

When collecting reptiles, it is worthwhile acquainting yourself with the proper techniques and equipping yourself with the correct apparatus.

If you turn over rocks or logs in search of specimens, please remember to replace them carefully in the original cavity or depression. Failure to do this can result in considerable damage to the environment or severe depletion of the habitat for restricted or rare species.

Equip yourself with stout gloves and a baling hook, and **see instructions for the avoidance of snake-bite on page 252.**

[1] *The National Parks and Wildlife Act 1974.*
[2] *The Wildlife Act 1975* — at October 1978 the Regulations pertaining to this Act are still in draft form, and until they are promulgated the Act has no force.
[3] *The Nature Conservation Ordinance* — the draft and Regulations passed the A.C.T. Legislative Assembly in June, 1978 and at October, 1978, is under consideration by the Federal Attorney General's Department for presentation to Parliament.

68

Collecting equipment

Noosing is a convenient and sure way of catching many lizards. Use a light but strong rod with a noose made of fishing line of selected gauge for the size of reptile to be caught. If you are cautious, most lizards will allow you to approach quite close. Slip the noose over the head of the animal and allow it to step through the loop with both forefeet before tightening. It is then usually a simple matter to transfer the captive animal to the collecting bag. The bag should be made from light canvas or calico and may be secured at the neck by tapes or cord.

For catching snakes, particularly the larger, venomous species, it is good practice to restrain the animal with a snake stick. The snake stick can be constructed from a stout pole with a Y-shaped stirrup tensioned across the Y by a two-centimetre wide leather strap and fixed at the end of the pole. The snake may thus be pinned to the ground by slight pressure across the neck region without causing injury to the specimen. The snake may then be picked up using a firm grip just behind the angles of the jaw and transferred, tail first, to the snake bag.

The safest type of snake bag should be constructed of moderately stout calico or light canvas with reinforced seams and a flexible wire loop inserted at the mouth. A short handle attached to this enables the mouth of the bag to be closed readily by simply turning the handle through 180°. The neck of the bag may then be secured with cord or strong tape which is most conveniently fixed to the bag by sewing it into position halfway along its length, leaving the ends free.

The bag may also be used to quieten the snake before capture. As most snakes become quiescent when placed in the dark and under such circumstances will form a compact coil, dropping the bag near or over the head of the snake will usually do the triek.

The dimensions of the bag may be determined from the size of the animal to be contained in it. It is usually better to allow a good depth, at least 60 centimetres, and about

35 centimetres width (when it is lying flat). When the animal is to remain in the bag for a long period, particularly when being transported, it is advisable to provide ventilation by means of reinforced eyelet holes let into the sides near the top of the bag and below the ties.

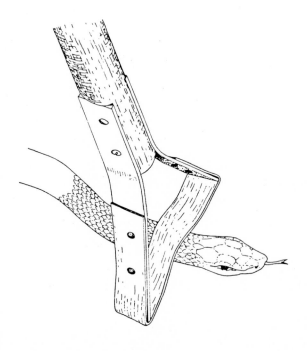

Figure 11
A snake stick. A leather strap is tensioned across the open end of a metal Y-piece and retained by rivets. The Y-piece is fixed to a wooden shaft about 1 metre long. If desired, an L-shaped metal 'jigger' may be fixed at the other end of the shaft for manipulating reptiles. This type of snake stick is superior to the all-metal T-piece type since there is less likelihood of the snake being injured when undue pressure is applied to restrain it.

Plastic bread boxes also make convenient collecting containers. A large window should be cut in the lid and a wire grid let in and fixed by heat-softening the plastic around the cut edge.

When on collecting trips always remember to place the captured specimens in as cool a place as possible. **Never leave live animals in a closed vehicle in the sun.**

Further information on collecting may be obtained from museums by writing to the Curator of Reptiles at State Museums where reptile collections are held.

Preservation of specimens

Animals which are destined for museum preservation may be dispatched using a 1:1 mixture of anaesthetic ether and chloroform. Museum specimens are fixed in 3% formalin, preserved in 70% ethanol (ethyl alcohol), and stored in tightly sealed glass containers. Dead specimens should be incised on the ventral surface of the body to allow the entry of preservative, but care should be taken to avoid the destruction or distortion of any diagnostic features while so doing. Fixative should also be injected into the tail using a hypodermic syringe.

Whenever specimens are presented to a museum it is essential that they be correctly, clearly and fully labelled if they are to be of any scientific use. Labelling should either be done at the time of capture or as soon as possible afterwards. Labels should never be gummed or tied to the outside of the receptacles containing the specimens, but should always be immersed in the preservative with the specimens. They can be tied to the specimen, and this will be necessary when more than one animal is placed in the same container. Labels should be of good quality paper and must be written in soft lead pencil, never ball-point pen. Ordinary inks will not withstand immersion, and the use of Indian ink is not recommended as only certain kinds are reliable.

The following particulars should always be entered on the label:

locality	—	(including, if possible, latitude and longitude) e.g. roadside, 31 km SSW Nerriga; 35° 22′ S. 150° 1′ E. (map references often help)
elevation	—	e.g. *c.* 700 m.
date	—	e.g. 23.xi.78
time	—	e.g. 15.30 hrs (Eastern Standard Time)
name of collector	—	
identified by	—	(name)
colour description	—	(colours often fade in preservative)
type of habitat	—	e.g. beneath stone in burrow, grassland

If collecting on a regular basis, it is advisable to keep a set of comprehensive notes entered under a collection number. Many useful details may not fit conveniently on to a label, so the note book entry and the label should be clearly marked with the same unique collection number or code. Sketches of the living or freshly dead animal, showing the distribution of body coloration and markings are also useful. It is of value to have some information regarding behaviour (if possible), whether eggs or live young were present, any associations with other animals, etc.

4
Descriptions and Plates

Notes on descriptions

In order to achieve some degree of clarity in the following section, a uniform format for the presentation of the descriptions has been maintained. If the reader is to be able to make full use of these descriptions, a number of aspects of the material and its arrangement need to be explained.

A reference to the type description accompanies each species prior to the full description; thus

'1838 *Tiliqua trilineata* Gray *Ann.Nat.Hist.***2**:291'.

It signifies that the type description of the Three-lined Skink *Leiolopisma trilineata* (Gray) was published in 1838 by Gray in the *Annals of Natural History*, volume 2, page 291. (See also page 59 in 3. Classification, Nomenclature and Diagnostics.)

Full synonymies for each species have not been given, principally because such an exercise would be of only marginal value in a 'popular' presentation designed to be used by both the professional herpetologist and the novice. In some instances, following the treatment of a species, there is a reference to the most recent relevant work on that species. Unfortunately not every species has been the subject of sufficient research to warrant this additional information. Where this is the case, the reader is directed to the recommended reading list at the end of the book (page 269).

Localities of the occurrence of each species (listed by State) are not intended to be exhaustive, even if that were to be possible in all cases. However, references to most known localities are given when museum specimens and their records have been examined, and the lists will provide an indication of the type of habitat in which each species is likely to be found.

It may be that the name applied to a locality is in fact some distance from the actual site where the specimen was collected, but it is the closest named reference point. For this reason, although the collection site lies within the boundaries of the Southern Highlands region, some named localities would appear to fall just outside. In some instances, however, the recorded site is immediately outside the defined region, and here the locality name is given in parentheses.

The body lengths quoted throughout the text, wherever possible, represent the average dimensions. However, in the case of many of the less common species this has not been possible. The dimensions given are in the following form:

$$T \ (sv \ + \ ot)$$

where T = total length in millimetres
         sv = snout to vent length in millimetres
         ot = original (non-regenerated) tail length in millimetres

In most instances, the figures quoted have been averaged from a large number of specimens. However, caution should be exercised when assessing the dimensions of rare or uncommon species as the figures for these have usually been obtained from only a few specimens and may represent an unduly biased estimate.

In cases where the local fauna includes more than one genus within a family, or more than one species within a genus, a key is provided to aid in identification. These keys are 'artificial' in the sense that they are constructed from characters useful in distinguishing individuals within the local taxa only, and may not be reliable for separating related taxa which are not represented in the Southern Highlands region.

Common names of species have been given only where they are in established use, or have been used extensively in the herpetological literature.

Finally, as it is not amongst the aims of this book to assess the evolutionary relationships between various species, rather than suggest or imply any order of evolutionary antiquity or phylogenetic affinities in the sequence of the species descriptions, we have chosen to deal with them in alphabetical order. This arrangement has been applied to each taxonomic level from order to species.

Aquatic tortoises: family *Chelidae*

The order Pleurodira contains all the *side-necked* tortoises. The two families, Pelomedusidae and Chelidae, contained within this order are restricted to the continents of the Southern Hemisphere. Chelid tortoises are found in equatorial South America, Trinidad, Argentina, Australia and New Guinea.

The family Chelidae is distinguished by the following features. The carapace and plastron are both constructed of bony plates with an overlying layer of horny shields. These are united on either side by a rigid bridge or similar structure. The eyes are situated towards the dorsal surface of the head and the nostrils are close to the top of the mouth. All the feet are fully webbed.

This family is represented in Australia and New Guinea by four genera which can be divided into two distinct groups. The first is distinguished by having exceptionally long necks; the genus *Chelodina*, of which there are six species, is the only representative of this division. In the second group, the seven species are characterised by having much shorter necks and they belong to the three genera *Elseya, Emydura* and *Pseudemydura*. The latter genus is monotypic and contains the rare, diminutive species *P. umbrina*. Another character which distinguishes the long-necked tortoises from their shorter-necked relatives is the condition of the digits. Long-necked tortoises have four clawed digits and one non-clawed digit on each limb; the short-necked tortoises have five clawed digits on the forefeet and four clawed digits and one non-clawed digit on the hindfeet.

All members of the family are oviparous.

Genus *Chelodina* Fitzinger 1826

Type species: *Testudo longicollis* Shaw 1802

Long-necked tortoises can be distinguished immediately from other Australian chelid tortoises by the possession of only four claws on webbed front feet. First vertebral shield longest. Pygal shield in contact with the caudal and four marginal shields, and not entirely covered by the last vertebral shield. Plastron anteriorly broad. Intergular shield large and surrounded by six other plates of the plastron. Skull long, being consistent with the elongate neck. Distributed throughout mainland Australia and New Guinea.

Chelodina longicollis (Shaw)   **Long-necked Tortoise** or **Snake-necked Tortoise**

1802   *Testudo longicollis* Shaw   *Gen.Zool.***3**:62

Localities   A.C.T.   Canberra, Bendora Dam, Mt Coree
           N.S.W.   Gundaroo, Bungendore, Collector, Murrumbidgee River and tributaries, Goulburn, Braidwood, Mongarlowe
           Vic.   (2.5 kilometres NE of Yea)

Diagnosis   The carapace (top shell) is brown and dorsally depressed. The plastron (bottom shell) is cream. The margins of the epidermal plates on the plastron are dark brown to black. The head and neck are elongate. All four limbs are well developed and pentadactyl. The forefeet are strongly webbed and each foot bears four claws, the fifth toe being clawless. The webbed hindfeet each bear five claws. There is a short postanal tail. The eyes and nostrils are set dorsally on the head. The nostrils are close to the tip of the snout. The skin is granular, and the scales are juxtaposed.

General biology   This tortoise is characterised by the long neck from which it derives its specific and common names. It appears to be most abundant in the lower altitude areas of the Southern Highlands, but it has also been recorded in some montane areas in the district.

Chelodina longicollis is semi-aquatic in habit. Its usual haunts are small water-holes and creeks where it remains submerged for extensive periods. On overcast days it may sometimes be encountered well away from free water. The Long-necked Tortoise is able to remain almost totally submerged in water. This is achieved by the dorsal displacement of the nostrils at the extremity of the head. When the animal is disturbed, the head may be withdrawn beneath the front margin of the carapace by swinging the neck sideways.

Food consists of aquatic arthropods, molluscs, frogs and plants.

Chelodina longicollis; Gungahlin A.C.T.
Photo courtesy of J. C. Wombey

The Long-necked Tortoise is oviparous, producing 10 to 15 small, ovoid eggs. The eggs are covered with a hard, white calcareous shell. Oviposition occurs during the summer months. The females take to the land where they find suitable locations for the excavation of the egg burrow. Incubation varies between two and three months, depending on soil temperatures. Hatching usually occurs between January and March.

Dragon lizards: family *Agamidae*

Agamid lizards are characterised by the rough appearance of the skin. The small, irregular scales which cover the head are unlike the plate-like scales of the skinks and most pygopodids. The scales of the body and limbs are imbricate and usually bear keels. The limbs are pentadactyl, and each digit bears a claw. The hindlimbs are often extremely long. The tail is long and thin, often being as much as three times the length of the body. Dragons do not have the trait of tail autotomy, but those species which have attenuated tails can dismember the end of the tail. There is usually either limited or no regeneration of the lost portion. The tympanum is distinct, except for the species of *Tympanocryptis* and *Amphibolurus maculosus* in which the ear opening is represented by a shallow depression. The teeth are situated on top of the jaw margins, the *acrodont* condition. The tongue is broad and fleshy. All dragons are oviparous.

There are eight genera of Australian agamids, containing such diverse forms as the Mountain Devil *Moloch horridus*, and the Frill-necked Dragon *Chlamydosaurus kingii*. The family is represented by at least one member in most climatic regions of Australia; the greatest degree of adaptive radiation has occurred within the genus *Amphibolurus*, particularly in the arid zones of the mainland interior.

Key to agamid genera

1 Tympanum exposed or, if covered by skin, all
dorsal scales are flat and smooth..(2)
No external ear opening, the tympanum being
covered by skin; some dorsal scales raised,
keeled, or spinose.. *Tympanocryptis*

2 Tail, at most, slightly laterally compressed
without a strongly-differentiated dorsal keel ....*Amphibolurus*
Tail strongly compressed with a strongly-
differentiated dorsal keel ....*Physignathus*

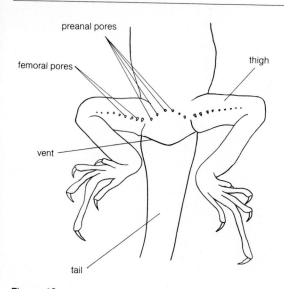

Figure 12
Diagram of the underside of an agamid lizard showing positions of the
preanal and femoral pores.

Genus *Amphibolurus* Wagler 1830

Type species: *Lacerta muricata* Shaw 1790

Head large. Tympanum distinct. Body dorso-ventrally compressed. Dorsal crest absent or feebly developed, gular sac absent, strong transverse gular fold present. Tail round, depressed or feebly compressed. Limbs pentadactyl and well developed. Preanal and femoral pores present. All are oviparous with terrestrial or arboreal habits. Distributed throughout mainland Australia and Tasmania.

Key to species

1 Buccal cavity coloured bright yellow .... (2)
   Buccal cavity flesh-coloured .... *diemensis*

2 Sides of base of tail with a series of spinose
   scales .... *barbatus*
   Sides of base of tail covered with scales
   which may be smooth or keeled, but never
   spinose .... *muricatus*

Amphibolurus barbatus (Cuvier)

**Bearded Dragon
or Jew Lizard**

1829 *Agama barbata* Cuvier   *Règne Anim.***3**:35

Localities A.C.T. Canberra and A.C.T. generally
            N.S.W. Yass, Goulburn, Gunning, Braidwood, Lake George, (Gundagai), (Cootamundra)
            Vic.   Bright

Diagnosis The dorsum is covered with keeled scales of varying sizes; there is an aggregation of large, prominently keeled scales in the vertebral region. The flanks are covered with several rows of small, extremely spinose scales which widen to a broad band immediately anterior to the hind limbs. The caudal scales are also heavily keeled. The head is angular in dorsal view. The spines on the head are moderately pronounced. The transverse row of spines on top of the head is curved and seldom contacts the lateral postorbital row of spines before turning posteriorly to meet the nuchal spines. There are four or five pairs of femoral pores and two or three pairs of preanal pores. The adpressed hindlimb barely reaches the shoulder. The 'beard' is raised by the hyoid bones of the lower jaw; a thin membrane covered with rows of spinose scales is supported by these bones. The beard is under muscular control, only becoming fully expanded laterally when the mouth is open. The buccal cavity is bright yellow. There is a distinct canthus rostralis.

The dorsal colour is usually a shade of brown or grey. Melanic forms are not uncommon. The flanks and the proximal regions of the limbs and tail are often suffused with bright yellow. The ventral surface is light, silvery-grey with a series of darker grey ocellations. Juveniles possess a distinct dorsal pattern. This pattern fades with age and is less pronounced in mature animals. The average length of an adult male is 443 millimetres (189 mm + 254 mm).

Amphibolurus barbatus; Canberra A.C.T.

General biology The Bearded Dragon is confined to the drier regions of the Southern Highlands, being a common inhabitant of the woodlands and dry sclerophyll forests.

This species is a true heliotherm, only becoming active during periods of higher temperature, when it may often be seen basking on fence posts and fallen timber. *A. barbatus* has semi-arboreal habits, being equally at home on trees or on the ground. The Bearded Dragon is not a particularly agile lizard, preferring to rely upon its cryptic coloration while remaining motionless on a log or tree stump, and only breaking its cover if approached too closely. Occasionally some individuals adopt a characteristic defensive attitude by opening the mouth to reveal the beard and bright yellow buccal cavity, and inflating the lungs and air sacs to increase the apparent size of the body (see page 45).

Amphibolurus barbatus is an oviparous species, producing 10 to 20 soft-shelled eggs. The gravid female excavates a hole in the soil prior to oviposition and, after the eggs have been laid, the hole is filled in. Incubation is achieved by solar heat and takes approximately three months.

This species is omnivorous, feeding on a wide range of invertebrates, small vertebrates and vegetable matter. The Bearded Dragon is often kept as a household or garden pet, adapting readily to captivity and eventually becoming quite tame.

BADHAM, J. A. (1976) 'The *Amphibolurus barbatus* species group (*Lacertilia:Agamidae*).' *Aust.J.Zool.***24**:423–43

Amphibolurus diemensis (Gray)   **Mountain Dragon**

1841 *Grammatophora muricata* var. *diemensis* Gray
Grey — *J.Exped.Disc.Aust.1837–39*

Localities A.C.T.   Piccadilly Circus, Bulls Head, Mt Coree, Brindabella Ranges

N.S.W.   Mt Kosciusko, Kiandra, Cabramurra, Adaminaby

Vic.   Mt Cobberas, Native Dog Plain, Walhalla, Mt Buffalo National Park, Mt St Bernard, Glen Wills, Mt Wellington, Dartmouth, Daveys Plains, Mt Hotham

Diagnosis   On both dorsal and lateral surfaces, there is a heterogeneous mixture of large spinose scales interspersed amongst the more numerous, small, keeled scales. The spinose scales are arranged in four longitudinal rows which extend beyond the vent on to the proximal region of the tail; they are randomly disposed on the flanks and appendages. The scales on the belly are also keeled. The head is moderately-sized relative to the body. The interorbitals number 11 to 13. There is a strong gular fold. The hindlimb when adpressed reaches the orbit.

Dorsally this species may be dark grey to almost black with a series of lighter-coloured vertebral blotches. Some individuals exhibit the striking colour combination of russet with light grey vertebral blotches. The belly may be either cream or light grey. The average length of adults is 155 millimetres (55 mm + 100 mm).

General biology   The Mountain Dragon is restricted to the high mountain areas of the Southern Highlands. It is the only Australian agamid to inhabit country which receives annual winter precipitation as snow. It occurs in the intermediate sclerophyll forests, sub-alpine and alpine communities of the mountains, and is extremely common in areas that have been

Amphibolurus diemensis; Brindabella Ranges A.C.T.

cleared for the construction of power lines, where there are quantities of fallen timber.

During the warmer months of the year it may be found basking on fallen timber or scuttling through the undergrowth in search of food, which consists of a wide range of small invertebrates. The Mountain Dragon does not show the speed and agility displayed by many of its inland relatives, and it is readily caught without too much effort. During winter this lizard may be found in an almost frozen condition beneath deeply buried logs and tree stumps.

Amphibolurus diemensis is oviparous, producing two soft-shelled eggs which are usually deposited beneath a rock or a log.

Amphibolurus muricatus (Shaw)

Jacky Lizard
or **Tree Dragon**

1790 *Lacerta muricata* Shaw   White — *J.Voy.N.S.W.*

Localities A.C.T. Orroral Valley, Kambah
         N.S.W. Gundaroo, Lake George, Goulburn, Sutton, Yass, Braidwood, Lake Bathurst, Gunning, Mt Kosciusko
         Vic.   Wulgulmerang, (Buchan Caves), Dargo, Shelley, Dartmouth, Benambra, (Healesville)

Diagnosis   The back is covered with a heterogeneous mixture of small, irregular, keeled scales and larger spinose scales. There is a small, barely-distinct dorsal crest formed by the spinose scales. The spinose scales are often arranged in four longitudinal lines. There are four preanal pores and three to four pairs of femoral pores; the femoral pores are restricted to the proximal region of the thighs. The tail is covered with strongly-keeled scales of varying sizes. The hindlimbs are long, reaching a point beyond the tympanum when adpressed.

Dorsally this species may be any shade of grey, brown or fawn with a series of yellowish, cream, light grey or light brown vertebral blotches, which in some individuals may coalesce to form two vertebral stripes. The belly varies from off-white to light grey. The buccal cavity and the tongue are bright yellow. The average length of adults is 307 millimetres (102 mm + 205 mm).

General biology   *Amphibolurus muricatus* is widespread throughout the Southern Highlands with the exception of the alpine regions. It inhabits dry sclerophyll forests and woodlands, where it is sympatric with the Bearded Dragon *A. barbatus*.

It is diurnal and semi-arboreal in its habits and is often seen sunning itself on a tree stump or log. When disturbed,

Amphibolurus muricatus; Timbillica State Forest via Eden N.S.W.
Photo courtesy of J. C. Wombey

this dragon retreats to the sanctuary of the tree-top, climbing with the aid of strongly clawed digits. Once in a tree, this species becomes extremely difficult to detect as its body pattern and colouring blend well with the colour of the bark. During the colder months of the year, this lizard may be found in a state of torpor under logs or rocks.

Amphibolurus muricatus is oviparous and the female deposits her eggs, which may number up to eight, in a burrow or depression which she excavates beneath a rock or log.

The lizard's diet consists primarily of insects, so this species does not usually do well in captivity.

Genus *Physignathus* Cuvier 1829

Type species: *Physignathus cocincinus* Cuvier 1829

Body laterally compressed. Nuchal and dorsal crests present as well as strong gular fold. Tympanum distinct. Limbs strong and pentadactyl, overlapping when adpressed; toes denticulated laterally. Preanal and femoral pores present; sometimes absent in females. All species oviparous. Distributed throughout continental Australia, New Guinea and the Indo-Malaysian Archipelago; represented in Australia by a single species.

Family: *Agamidae*

Physignathus lesueurii howittii[1] McCoy   **Gippsland Water Dragon**

1878 *Physignathus lesueurii howittii* McCoy
Proc.Zool.Soc.Victoria, 1:7–10

Localities A.C.T.   Tidbinbilla Nature Reserve, Condor Creek, Murrumbidgee River and tributaries, Lake Burley Griffin (Canberra)

N.S.W.   Wee Jasper, Tarago, Jugiong, Burra, Goodradigbee River, Braidwood, Kosciusko National Park

Vic.   Suggan Buggan, Gelantipy, Buchan River (upper reaches), junction of Snowy and Broadbent Rivers, Chandlers Creek, Omeo

Diagnosis   The head is large and angular when viewed from above. The tympanum and canthus rostralis are both distinct. There is a heterogeneous arrangement of the dorsal scales. Large heavily-keeled scales are randomly scattered amongst the more numerous small unicarinate scales. A dorsal crest extends from a point midway between the tympanic depressions to the tip of the tail, being most prominent in the nuchal region. The limbs are well developed and pentadactyl. The hindlimb almost reaches the tympanum when adpressed. There is a strong transverse gular fold. The tail is laterally compressed in the distal region.

Dorsally the male is green with a bluish tinge. There is a series of black transverse bars which extend the entire length of the body and tail. The ventral surface is light grey, mottled with black. The gular region of mature males is most strikingly coloured, being a reticulated pattern of yellow and blue. The females and juveniles are drab brown with poorly-defined

[1]There are two recognisable races of this species: *'howittii'* — referred to here, and the nominate race *'lesueurii'* (Eastern Water Dragon) which inhabits riverine environments of lower elevations and coastal eastern Australia.

black transverse bands. The average length of adult males is 730 millimetres (250 mm + 480 mm); the females are smaller.

General biology The Gippsland Water Dragon, as its name suggests, lives in close association with waterways. During the summer months, it is frequently to be found basking on rocks and timber adjacent to watercourses in the lower elevations on the Southern Highlands, and is often present in quite large numbers.

Physignathus howittii is a shy lizard and consequently can be approached only with considerable difficulty. When disturbed this lizard takes to the water very quickly, running in bipedal fashion; once in the water, it submerges and swims to cover. Swimming is accomplished by adpressing both forelimbs and hindlimbs to the body, and throwing the body into a series of undulating curves. It is able to remain submerged for considerable periods of time. The Gippsland Water Dragon is an adept climber, as it is well endowed with strongly-clawed digits. It may often be encountered basking in trees along the margins of rivers, from which position it can dive into the safety of the water when disturbed.

The diet of this species consists of invertebrates, amphibians, small reptiles and mammals. On one occasion, a large male was observed beside a river feeding on a large Golden Carp *Carassius auratus* (Jenkins, personal observation).

The Gippsland Water Dragon is an egg-laying species, producing approximately eight soft-shelled eggs which are deposited either under a rock or at the end of a burrow dug by the female. The period of development of the eggs varies from 10 to 14 weeks, depending on soil temperature.

Physignathus lesueurii howittii; Cotter River, A.C.T.

Genus *Tympanocryptis* Peters 1863

Type species (by monotypy): *Tympanocryptis lineata* Peters 1863

Small, terrestrial, cryptozoic lizards with a depressed head, body and tail. External auditory meatus absent. Strong gular fold present. Preanal pores usually two, but absent from females of most species. All species oviparous. Distributed throughout continental Australia with the exception of the east coast and south-western Australia.

Family: *Agamidae*

Tympanocryptis lineata Peters   **Earless Dragon**

1863 *Tympanocryptis lineata* Peters

*Monatsber.K.Preuss.Akad.Wiss.Berlin,***230**

Localities   A.C.T.   Canberra
            N.S.W.   Cooma
            Vic.   nil

Diagnosis   The dorsum is covered with large spinose tuber-cules interspersed amongst the more-numerous small, smooth, irregular scales. The spinose scales appear to be closely asso-ciated with the pigmented areas. The belly scales are smooth. The hindlimbs are well developed, and will reach the promi-nent transverse gular fold when adpressed. The tail is swollen at the base. There is a pronounced canthal ridge. The nostril is pierced in a large nasal. There are two preanal pores present in both sexes. The interorbitals average 18. There are seven to nine rows of scales between the orbit and the lip.

The basic dorsal colour is light grey. Five brown transverse bars extend between the head and the hindlimbs and there are numerous transverse bars along the tail, becoming indis-tinct in the distal region; the bars are interrupted by five white longitudinal lines commencing on the neck. The vertebral line extends to the pelvic region. The two dorso-lateral lines are confined to the region between the appendages. The head is mottled. The ventral surface of the female is white with fine black flecks, which are most concentrated in the gular region. The belly of the male is immaculate white. During the breed-ing season the gular region of the male becomes bright yellow whilst the pelvic region becomes distinctly pink. The average length of adults is 80 millimetres (35 mm + 45 mm).

This lizard was recognised by Mitchell (1948) as a distinct eastern race, *pinguicola*, occurring in south-eastern New South Wales and eastern Victoria. We have decided to include this form in the nominate race *lineata*, as there is a continuous distribution of this race from the Nullabor region in Western

Australia, through South Australia and Victoria to the south-eastern highlands. Mitchell's major diagnostic used in separating the *pinguicola* race from the nominate race, apart from the apparent geographical isolation, was the relative widths of the head and neck. We have found that, in practice, this character is generally inconsistent and therefore of doubtful taxonomic value.

General biology   This small agile agamid has a very limited distribution in the Southern Highlands region. *T. lineata* is the only species of this genus which extends onto the Great Dividing Range of eastern Australia. The genus *Tympanocryptis* has achieved its greatest radiation in the arid zones of central and Western Australia. On the Southern Highlands, it appears to be restricted to the warmer regions, inhabiting the grasslands and savannah woodlands. Even within these types of habitat, the species is extremely rare and few specimens have been recorded from the area.

Tympanocryptis lineata is diurnal and wholly terrestrial in its habits. During the summer months it may be encountered scuttling through the grass or sunning itself on a log or rock. When disturbed, it retreats rapidly into the confines of grass tussocks where it is extremely difficult to detect. During winter months it may be found in a torpid state under rocks in a shallow depression. In several instances, specimens have been found utilising the abandoned burrows of Trap-door Spiders of the family Lycosidae.

The diet of this lizard consists principally of small insects and other terrestrial invertebrates. This species is oviparous, although nothing is known of the egg clutch size or oviposition site.

MITCHELL, F. J. (1948) 'A revision of the lacertilian genus *Tympanocryptis.*' *Rec.S.Aust.Mus.***9**:57–86
STORR, G. M. (1964) 'The agamid lizards of the genus *Tympanocryptis* in Western Australia.' *J.Proc.R.Soc.West.Aust.***47**:43–50

Tympanocryptis lineata pinguicola; Victoria
Photo courtesy of F. Collet

Geckoes: family *Gekkonidae*

The family Gekkonidae is found in the warmer regions of both the Old and New Worlds. It is well represented in Australia with more than a dozen genera.

Geckoes are small to moderate-sized nocturnal lizards. They are characterised by the possession of large eyes with elliptical pupils; a flabby skin covered with small, usually juxtaposed, granular scales; eyes covered with a transparent scale, the lower eyelid being immovable; a distinct tympanum; and four well-developed pentadactyl limbs.

The condition of the digits is variable. Geckoes which are principally terrestrial — *Nephrurus, Heteronotia, Cyrtodactylus, Phyllurus* and *Rhynchoedura* — have simple, clawed digits. Those forms which have arboreal habits — *Gehyra, Oedura* and *Phyllodactylus* — have digits which are dilated at the apices. The dilation is formed by rows of subdigital lamellae which give purchase on smooth surfaces. The degree of dilation and its relative elaboration is exemplified by members of the genus *Diplodactylus*; this is the largest and most diverse gekkonid genus in Australia, containing arboreal, fossorial and ground-dwelling species.

Geckoes will readily dismember the tail when handled; the lost portion is regenerated but will never redevelop to the original extent. Some species are able to vocalise quite loudly.

All members of the family are oviparous and most species lay a clutch of two eggs. All species produce hard, calcified-shelled eggs.

Key to gekkonid genera

Scales above distal expansions of digits significantly larger than scales above basal parts of digits.. *Phyllodactylus*

Scales above distal expansions of digits more or less equal in size to those above basal parts of digits..*Diplodactylus*

Genus *Diplodactylus* Gray 1832

Type species: *Diplodactylus vittatus* Gray 1832

Digits not dilated at the base. Distal expansion covered above with small scales similar to those scales on the basal part. Retractile claws arising between the two plates under the extremity of each digit. Labials larger than or equal to adjacent scales. Scales on the upper and lower surfaces of the body generally juxtaposed. Oviparous. Terrestrial and arboreal geckoes distributed throughout continental Australia.

Diplodactylus vittatus Gray

Stone Gecko

1832 *Diplodactylus vittatus* Gray

*Proc.Zool.Soc.London,***1832**(2):40

Localities A.C.T. Mt Ainslie, Mt Majura, Black Mountain, Coppins Crossing

N.S.W. Lake George, Gundaroo, Bungendore, Yass, Goulburn, Cooma

Vic. Beechworth, Tintaldra.

Diagnosis The body is covered with small juxtaposed, granular scales. There is only a slight dorso-ventral differentiation of scales, those on the dorsal surface being slightly larger. The scales become slightly larger on the tail. Below each digit there is a series of enlarged plate-like tubercles. The distal region of each digit is made up of two ventral plates, producing a slight apical dilation, and the dorsal surface is covered with small granular scales. A small retractile claw is situated at the extremity of each digit. The pupil is elliptical. The interorbital scales number 26 to 38. The rostral is rounded, and is concentric with the contour of the snout. The anterior nasal is absent. The first supralabial borders the margin of the nostril. There are no preanal pores.

The dorsal surface may be brown or grey. There is a lighter-coloured, zig-zagged vertebral stripe which bifurcates in the region of the neck, forming a light-coloured crown. The ventral surface is light grey. The tail is short and thick, tapering to a point. The average length of adults is 85 millimetres (55 mm + 30 mm).

General biology This small robust gecko is terrestrial in habit, living under rocks and logs in well-drained areas. It inhabits dry sclerophyll forests and savannah woodlands and is widespread throughout Australia. The Stone Gecko is a nocturnal lizard, as are all other members of the family.

Like all geckoes, *D. vittatus* is able to vocalise during times of stress. When aroused, it raises its whole body off the ground

101

and, with its mouth agape, makes short lunges towards its tormentor. The Stone Gecko is able to dismember its tail voluntarily if alarmed or restrained by the tail, but the regenerated portion does not retain the original body pattern and is invariably shorter than the original.

The diet consists mainly of insects and other small arthropods which are caught at night.

It is an oviparous species. The eggs have calcified shells and there are usually two. The oviposition site is commonly under a rock or log.

KLUGE, A. G. (1967) 'Systematics and zoogeography of the lizard genus *Diplodactylus* Gray (Gekkonidae).' *Aust.J.Zool.* **15**:1007–1108

Diplodactylus vittatus; Mt Ainslie A.C.T.

Genus *Phyllodactylus* Gray 1828

Type species (by monotypy): *Phyllodactylus pulcher* Gray 1828

Digits dilated at the apex, with two large plates below; scales on the upper surface of the dilation differ in size from those covering the basal region. All digits possess a retractile claw. Rounded rostral and mental shields. Enlarged labials. Pupil is vertical. Dorsal scales juxtaposed, ventral scales imbricate. Femoral and preanal pores absent. Oviparous. Distributed from America through Africa, Mediterranean islands and Asia to Australia and its external territories; absent from arid areas.

Phyllodactylus marmoratus (Gray)   **Marbled Gecko**

1844 *Diplodactylus marmoratus* Gray
           Zool.Voy. Erebus *and* Terror *Rept.*Pl.*xv*,Fig.6

Localities   A.C.T.   Black Mountain, Mt Painter, Mt Ainslie, Mt Majura

                N.S.W.   Sutton, Gundaroo, Coolac, Goulburn, (Jugiong), (Holbrook)

                Vic.   Strathbogie Ranges, (Yea)

Diagnosis   The digits are dilated at the apex. A poorly-developed claw arises on the dorsal surface of each digit between the two plates which form the dilation. There is a single row of subdigital lamellae. The scales above the distal expansion of the digits are larger than those above the basal region of the digits. The dorsal body surface is covered with small juxtaposed, granular scales. The scales on the ventral surface are larger and imbricate. The caudal scales are arranged in annular series. There are neither preanal nor femoral pores. The pupil is vertical.

    The dorsal surfaces of the body and tail are grey with dark or black reticulations. The ventral surface is light grey. The average length of adults is 100 millimetres (45 mm + 55 mm).

General biology   Although this small gecko is widely distributed throughout the warmer regions of the Southern Highlands, it cannot be regarded as being common in any particular habitat type. It inhabits dry sclerophyll forests and savannah woodlands, sometimes being found in sympatry with the Stone Gecko, *Diplodactylus vittatus.*

    It exploits such microhabitats as leaf litter, decorticating bark and exfoliating rock. The general body shape is flattened dorso-ventrally which enables this species to exploit fully the narrow cracks in rocks or under bark, either in search of food or to escape predation. The Marbled Gecko is able voluntarily to shed its tail and regenerate a new portion. However, as is the case with *D. vittatus*, the original body pattern is not reproduced on the new tail. It is a nocturnal species and individuals may be encountered on warm summer nights running over tree trunks; an activity which they are able to perform with the aid of 'adhesive' plates on the under-surface of the feet.

Phyllodactylus marmoratus is oviparous and lays two to three hard-shelled eggs which are usually deposited under objects on the ground. Its diet is similar to that of the Stone Gecko and consists of insects and other small invertebrates.

Phyllodactylus marmoratus; Mt Ainslie A.C.T.

Flap-footed lizards: family *Pygopodidae*

The family Pygopodidae is restricted to Australia and New Guinea. More than 24 species are represented in eight genera. All but the two species of *Lialis* occur exclusively in Australia.

All members of the family are characterised by their body shape which is elongated and snakelike. The forelimbs are completely absent, whilst the hindlimbs are reduced to two lateral flaps; in the fossorial species of *Aprasia* they are barely visible. Dorsally the head is covered with numerous regular shields, except in *Lialis*. The tympanum is distinct, except in *Aprasia*. Preanal pores may or may not be present. The body scales are imbricate and may be either smooth or keeled. The ventral scale rows are only marginally wider than the other body scales, and never extend across the body as in snakes. In all cases, except *Aprasia*, the tail (when normal) is attenuated and longer than the snout to vent length; such tails are fragile and easily regenerated after autotomy. Species of *Aprasia* have a short, blunt tail. The eye is covered with a single, fused, transparent scale. The tongue is broad, fleshy and feebly bifid at the tip.

Pygopodids can easily be separated from degenerate-limbed skinks by one or more of the following characters: the absence of movable eyelids; possible presence of preanal pores; and flap-like condition of the hindlimbs.

All pygopodids are oviparous.

KLUGE, A. G. (1974) 'A taxonomic revision of the lizard family Pygopodidae.' *Misc.Publs.Mus.Zool.Univ.Mich.***147**:221 pp.

Key to pygopodid genera

1 Head covered with enlarged, symmetrical
  plates...(2)
  Head covered with small irregular shields..................................*Lialis*

2 Preanal pores absent...(3)
  Preanal pores present.. *Pygopus*

3 Parietal scales present...*Delma*
  Parietal scales absent..*Aprasia*

Genus *Aprasia* Gray 1839

Type species (by monotypy): *Aprasia pulchella* Gray 1839

Small species characterised by the absence of the parietal shields and lack of preanal pores. Head and body of equal diameters; head covered with symmetrical plates. Snout is blunt. Tympanum indistinct. Postanal tail short and blunt. Vermiform habit. Distributed throughout mainland Australia; absent from Tasmania.

Family: *Pygopodidae*

Aprasia parapulchella Kluge

1974 *Aprasia parapulchella* Kluge
                    *Misc.Publ.Mus.Zool.Univ.Mich.***147**:221pp.

Localities A.C.T.   Coppins Crossing (type locality), Black
                          Mountain
                N.S.W.   (Cootamundra), (Tarcutta)
                Vic.   nil

Diagnosis   This species is slenderly built. There are no parietal shields. The nasal and the first supralabial are fused.
It is characterised by the presence of a single postocular and
usually two preocular scales. There is only one elongate
supraocular. The anal flaps are indistinct. There is neither
an external auditory meatus nor preanal pores. There are
three enlarged preanal scales, two or three preorbital scales
and two postorbital scales. The body scales are smooth and
in 14 rows. The lower jaw is considerably undershot.

The general body colour is light grey with a series of fine
dark longitudinal lines which extend the entire length of the
body, becoming confused on the tail. The belly is a lighter
shade of the dorsal colour.The short, blunt tail is either cream
or pink. The average length of adults is 180 millimetres
(105 mm + 75 mm).

General biology   *Aprasia parapulchella* appears to have a limited distribution on the Southern Highlands. However, as the
habitat types of the localities given above are by no means
restricted in the region, it is likely that further field work will
reveal a considerably wider distribution.

The species inhabits well-drained, granitic country where
it leads a fossorial existence. It may be found beneath slabs
of rock which have not penetrated the soil surface very deeply.
It is a gregarious species; several individuals may be found
beneath the same rock. *A. parapulchella* is an adept burrower
and is extremely difficult to track down if it is not caught
as soon as it is uncovered. The species is nocturnal.

Aprasia parapulchella; Coppins Crossing A.C.T.

Reproduction is oviparous. One female collected in December had two well-developed, elongate ova inside her.

The diet consists of small arthropods. An analysis of the gut contents of five specimens revealed that small ants of the genus *Iridomyrmex* are the most usual food. In fact, most of the specimens found in the field have been under rocks that harbour colonies of these ants. Even though the postanal tail is very much shorter than the snout-vent length, this species is still able to shed its tail voluntarily when necessary.

Genus *Delma* Gray 1831

Type species (by monotypy): *Delma fraseri* Gray 1831

Moderate-sized species with smooth, imbricate body scales. The two median ventral series and the single subcaudal transversely enlarged. Head somewhat tapering, snout rounded, covered with symmetrical plates; frontal and prefrontal of approximately equal size. Tympanum distinct. Preanal pores absent. Vestiges of hindlimbs well developed. Postanal tail considerably longer than snout-vent length. Oviparous. Australia-wide distribution with the exception of Tasmania.

Key to species

Stripes absent on body and tail; nasal and first
supralabial not fused anterior to nostril .... *inornata*
One or two narrow dorsolateral stripes on body
and tail; nasal and first supralabial fused anterior
to nostril .... *impar*

Delma impar (Fischer)

1882 *Pseudodelma impar* Fischer   *Arch.Naturgesch.***48**:287 P1.16

Localities A.C.T.   Canberra, Gungahlin, Black Mountain, Barton

          N.S.W.   Sutton, Tumut, Gilmore, (Tarcutta)

          Vic.   (Yea)

Diagnosis Head covered dorsally with large symmetrical plates. There are usually five or six loreal scales and five anterior orbital scales. The fourth supralabial is subocular. There are usually four frontal scales which are made up of an enlarged pair of frontals (immediately posterior to the nasals), a single prefrontal scale and a single large frontal. Following the frontal is a pair of large parietals, which is followed medially by a distinct nuchal scale. The major diagnostic used by Kluge (1974) in distinguishing this species is the partly-fused condition of the first supralabial and the nasal anterior to the nostril. However, two specimens which were collected from the Australian Capital Territory possessed a complete suture between the nostril and rostral. The eye is not completely surrounded by small palpebral scales, as the dorsal margin of the orbit is covered by an elongate supraciliary scale. There are 15 to 16 rows of smooth scales around the body with a paired series of enlarged ventrals, the average number being 65.7 in males and 70.5 in females. There are two enlarged preanal scales.

Dorsally *D. impar* is olive green in colour. The lateral body colour is chocolate brown, varying from a broad band four scales wide to two narrow stripes each half a scale wide. Specimens which display the latter condition also have bright salmon-coloured flanks. The stripes become disrupted posterior to the vent and give the tail a reticulated appearance. The throat and belly are immaculate cream. This species is one of the smallest members of the genus, the average length of adults being 150 millimetres (60 mm + 90 mm).

112

Delma impar; Canberra A.C.T.

General biology   This species does not appear to be a common inhabitant of the Southern Highlands region; its known range is extremely limited. It occurs beneath rocks and debris in grasslands, woodlands and dry sclerophyll forests.

Delma impar is nocturnal in habit, and is seen to be active well after sundown during the warmer periods of summer. On overcast cooler days, its activity cycle is not nearly so well defined and could best be described as discontinuously diurnal and crepuscular. On a number of occasions, individuals have been unearthed by gardeners when they were found to be entangled in the roots of shrubs.

The locomotory behaviour of *D. impar* is extremely snake-like, as the body is thrown into a series of sinuous undulations. However, the species' capacity for controlled movement is not as well developed as it is in snakes, particularly in specimens that are devoid of the greater part of the tail through autotomy. When handled roughly, *D. impar* twists violently, an action which often leads to voluntary tail-breakage. Accompanying this behaviour the animal emits audible squeaks and spreads out the flap-like rudiments of the hindlimbs.

The diet of this species consists of small arthropods. The species is oviparous, but nothing is known of the oviposition site and the size of the egg clutch.

Delma inornata Kluge

1974 *Delma inornata* Kluge
Misc.Publ.Mus.Zool.Univ.Mich.**147**:221 pp.

Localities A.C.T.   Kambah Pool, Canberra, Pine Island
N.S.W.   Murrumbateman, Yass, Dog Trap Ford,
Gundagai
Vic.   Bright

Diagnosis The head is covered with large symmetrical plates; the nostrils are pierced in a pair of nasal scales, which are medially in contact immediately posterior to the rostral. There are four frontonasals (2 + 2), the posterior pair of which are enlarged and bordered posteriorly by a single prefrontal. The single frontal is followed by a pair of large parietals. *D. inornata* does not possess a distinct nuchal scale. There are six supralabials of which the fourth is elongate and subocular. The eye is completely surrounded by uniformly-sized palpebral scales. There is a series of four supraoculars. The gular scales usually number 14. The posterior nasal is widely separated from the nostril. The body scales are smooth and in 16 rows around the body with a pair of enlarged ventral scales. The tympanum is distinct and oblique. The tail is extremely long and attenuated.

The colour dorsally is olive green. There is an indistinct reticulated pattern produced by the black anterior margins of some of the body scales. The throat and ventral surface are flesh-coloured and devoid of any markings. The gular region is often bright yellow. The ocular scales are immaculate white. There is no apparent sexual dimorphism. The average length of adults is 350 millimetres (100 mm + 250 mm).

General biology *Delma inornata* is a moderate-sized species which inhabits the open savannah woodlands. It is possible that it has a wider distribution in the Southern Highlands region than is generally thought as the type of country which harbours this species is quite extensive throughout the lower

Delma inornata; Yass N.S.W.

altitudes of the region, particularly on the western slopes. It is doubtful, however, whether it could be regarded as being common within its range. It is absent from the montane regions.

Delma inornata is a terrestrial species with crepuscular habits. It may be encountered during the warmer months in grass around rocky outcrops or fallen timber. During winter, this species may be found in a torpid state either beneath or within rotten logs and under rocks. In the latter situation, it may be found in the company of *Carlia tetradactyla* and other lizards of the sub-family Lygosominae, which also constitute the bulk of its diet.

The species is oviparous, but nothing is known of the size of the egg clutch or oviposition site.

Genus *Lialis* Gray 1834

Type species (by monotypy): *Lialis burtonis* Gray 1834

Large, thick-bodied species. Pointed head covered with small, irregular scales. Body scales smooth, pointed and imbricate. Vestiges of hindlimbs barely discernible. Four preanal pores present. Postanal tail much longer than snout-vent length. Oviparous. Distributed throughout Australia, with the exception of Tasmania, and New Guinea.

Lialis burtonis Gray

Burton's Legless Lizard

1834 *Lialis burtonis* Gray

Proc.Zool.Soc.London, **1834**:134

Localities A.C.T.   Canberra
              N.S.W.   Tumbarumba
              Vic.   nil

Diagnosis *Lialis burtonis* is a thickly-built species. The mid-body scales are smooth and in 19 to 21 rows; two rows of ventral scales are enlarged to form plate-like ventrals. There is a distinct canthus rostralis. The head is elongate, tapering to a sharp snout. The nostrils are small and are pierced through reduced nasals. The pupils are elliptical. The head is covered with small irregular scales. The hindlimbs are indistinct. There are three to five enlarged anal scales preceded by four preanal pores.

The coloration of this species is variable. The A.C.T. specimen was uniformly light grey with a faint white line passing through the lips, along the body wall, and becoming indistinct anterior to the vent. The average length of adults is 520 millimetres (240 mm + 280 mm).

General biology Only one specimen has been recorded in the Southern Highlands during the preparation of this book. It was found in dry sclerophyll forest where it was disturbed whilst sunning itself in the undergrowth. This species is absent from the montane regions of the Southern Highlands, being confined to the lower altitudes where the prevailing climatic conditions are not so severe. *L. burtonis* is one of the most widespread of Australian reptiles occurring throughout the continent, except Tasmania, from high-rainfall coastal areas to the deserts of the inland.

As would be expected with a reptile that inhabits a wide range of habitat types, this lizard occurs in a number of colour varieties which range from pink or yellow to grey or dark

117

Lialis burtonis

brown. Some individuals display the rare reptilian character of a ventral surface that is darker than the dorsal surface.

Lialis burtonis is an oviparous species, producing two to three elongate, soft-shelled eggs which are deposited under a rock or a log. The Southern Highlands specimens are predominantly diurnal in habit, although individuals which inhabit the tropical regions are necessarily nocturnal for most seasons of the year. This species, like all pygopodids, is able to shed its tail voluntarily if the animal is restrained by it. It can also become quite vocal when handled roughly, uttering a series of loud squeaks. Its diet consists of insects, small skinks and geckoes.

KLUGE, A. G. (1974) 'A taxonomic revision of the lizard family Pygopodidae.' *Misc.Publs.Mus.Zool.Univ.Mich.* **147**:221 pp.

Genus *Pygopus* Merrem 1820

Type species: *Bipes lepidopodus* Lacépède 1804

Moderate to large, robustly-built species. Head covered with large symmetrical plates and small scales; frontal much larger than the prefrontal. Tympanum distinct. Body scales hexagonal and imbricate; the two central rows of abdominal scales transversely enlarged. The rudiments of the hindlimbs are prominent and paddle-shaped. Preanal pores present. Tail much longer than head and body. Oviparous. Distributed throughout mainland Australia, with the exception of the alpine regions of the south east.

Pygopus lepidopodus (Lacépède)

Common Scaly Foot

1804 *Bipes lepidopodus* Lacépède
Ann.Mus.d'Hist.Nat.Paris **4(193)(209):**P1.*lv,* Fig. 1

Localities A.C.T. Tidbinbilla Nature Reserve
 N.S.W. Cabramurra
 Vic. nil

Diagnosis This large heavily-built species is identified by the presence of large symmetrical plates on top of the head, and a series of preanal pores, which number 10 to 14. The head is large and not compressed; the snout is rounded. The plates of the head are arranged in the following series: the nasals make contact posterior to the rostral; four enlarged frontonasals $(2+2)$ make broad contact anterior to a single prefrontal; the frontal is followed by an elongated pair of broadly-contacting parietals. There are seven supralabials, the fourth being subocular. The tongue is broad, fleshy and feebly bifid at the tip. There are 68 to 88 pairs of ventrals which are transversely larger than the adjacent body scales. The paddle-shaped hindlimbs are proportionately longer than in other pygopodids, being five scale rows in depth and seven in length. There are 20 to 24 scales around the body. The body scales are imbricate and sharply keeled, and increase in size towards the ventral surface. The keeled condition extends on to the caudal scales. The tympanum is distinct and oblique. There are three enlarged anal scales. The tail is extremely long, round in cross section, and tapers to a point.

 Dorsally this species is uniformly coloured, varying from grey to olive green or reddish brown. The ventral surface varies from light grey to cream or pink. There are a number of black mottlings on the belly, becoming confused on the tail. There is often a series of black markings associated with the lips. The average length of adults is 470 millimetres (150 mm + 320 mm).

General biology   *Pygopus lepidopodus* is the largest member of the family Pygopodidae, and can attain a length greater than 600 millimetres. Although only one specimen has been recorded from the Southern Highlands region; the habitat in which it was found is not unique to this one locality. It is reasonable to assume that more intensive fieldwork would reveal a considerably wider distribution, particularly in the northern sector. The species can be expected to inhabit the grasslands, woodlands and dry sclerophyll forests of the lower elevations where it exists under ground debris such as rocks and logs. It is absent from the montane areas.

On the Southern Highlands *P. lepidopodus* is diurnal, but in the warmer regions of its range it is known to be nocturnal in its habits. When aroused, this species displays the habit of raising the anterior regions of its body well off the ground and compressing the neck region in much the same manner as the Brown Snake. This behaviour combined with the

Pygopus lepidopodus; Bulli N.S.W.

lizard's snake-like form serve as effective deterrents to would-be predators. If handled roughly, it is able to break its tail voluntarily. The lost portion is later imperfectly regenerated.

The species is oviparous and produces two large elongate, soft-shelled eggs, which are deposited beneath a rock or log in mid-summer. Nothing is known of the development period of the eggs.

The diet of *P. lepidopodus* in this region consists predominantly of small lygosomine skinks of the genus *Leiolopisma*.

Skinks: family *Scincidae*

This family has a world-wide distribution; its name is derived from the type genus *Scincus*. Members of the family are poorly represented in America and Europe, but are extremely abundant and diverse in Africa and South East Asia. In Australia, they comprise the greater proportion of endemic lizards, and occupy a wide range of habitats.

Skinks can be distinguished readily from the geckoes, dragons and monitors by the large plates on the top of the head and their imbricate body scales. The dentition is pleurodont. The tongue is moderately long, free and feebly bifid at the tip. The limbs are variable, and may be of moderate length, short or even absent. There are usually five fingers and toes, but there may be fewer in some groups; the lower number of digits is usually associated with limb reduction. The nature of the tail is also variable; it may be long and slender, short or even stumpy. The potential for tail autotomy is most commonly found in the species that possess long, attenuated tails.

Australian skinks display one of the following three conditions of the eyelids: lower eyelid movable but scaly and opaque, e.g. *Tiliqua* and *Egernia*; or lower eyelid movable and containing an undivided, transparent palpebral disc, e.g. *Leiolopisma* and *Carlia*; or lower eyelid fused and immovable, the eye covered with a transparent scale, e.g. *Cryptoblepharus*.

Members of this family vary in length from several centimetres to more than a metre. The three modes of reproduction, oviparity, ovoviviparity and viviparity, are all represented in the skinks.

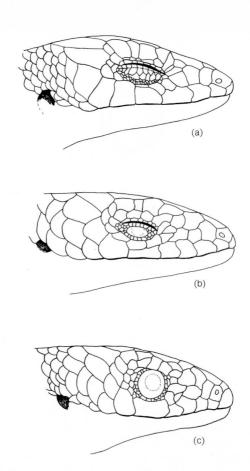

Figure 13
Three conditions of the lower eyelid of skinks.
(a) Lower eyelid movable and covered with small, delicate palpebral scales (*Sphenomorphus*)
(b) Lower eyelid movable with a large transparent window — the palpebral disc (*Leiolopisma*)
(c) Lower eyelid fused and immovable, forming a transparent spectacle covering the eye — the 'ablepharine' condition (*Morethia*)

124

Key to scincid genera

1   Parietal shields (if unfragmented) are not in contact, being separated behind the interparietal .... (2)
     Parietal shields in contact behind the interparietal .... (4)

2   Third and fourth toes more or less equal or the third slightly longer than the fourth .... (3)
     Fourth toe markedly longer than the third .... *Egernia*

3   Tail moderate to long, tapering; dorsal scales of moderate size, smooth; head shields smooth, entire and symmetrical; subdigital lamellae undivided .... *Tiliqua*
     Tail short, depressed, blunt-ended; dorsal scales grossly enlarged, strongly rugose; head shields fragmented, only vaguely symmetrical; subdigital divided (at least basally) .... *Trachydosaurus*

4   Lower eyelid with transparent disc and either movable or fused to form a transparent spectacle .... (5)
     Lower eyelid movable, scaly, or with an opaque disc .... (11)

5   Lower eyelid movable, partly or totally fused to form a permanent spectacle; if the lower eyelid is fused, the prefrontals are small and widely separated, or absent .... (6)
     Lower eyelid totally fused to form a permanent spectacle; prefrontals large, in contact or narrowly separated .... (10)

6   Supranasals absent and nasals undivided .... (7)
     Supranasals present, or nasals divided .... *Pseudemoia*

7   Limbs well developed, meeting or overlapping when adpressed, or else separated by one or two scale lengths; ear opening prominent .... (8)
     Limbs short, separated by at least several scale lengths when adpressed; ear opening small to minute or hidden .... (9)

8 Fingers four, toes five .... *Carlia*
   Fingers and toes five .... *Leiolopisma*

9 Nasals small to moderate, usually separated .... *Hemiergis*
   Nasals enlarged, usually in contact medially .... *Lerista*

10 Fingers and toes five .... *Morethia*
   Fingers four, toes five .... *Menetia*

11 Conspicuous ear lobules present; pattern usually of dorsal and/or lateral longitudinal stripes .... *Ctenotus*
   Ear lobules absent; pattern usually transversally aligned, or of irregularly scattered spots and variegations .... *Sphenomorphus*

Genus *Carlia* Gray 1845

Type species (by monotypy): *Mocoa melanopogon* Gray 1845

Small, stoutly-built skinks. Head sub-quadrangular; rostral erect, triangular or convex; nasals lateral, nearly contiguous; supranasals absent; palate toothless; auditory meatus oblong, anteriorly denticulated. In all but one species (*C. burnettii*) the lower eyelid is movable, with a transparent palpebral disc. Body fusiform. Limbs four, moderately built; fingers four, toes five, all compressed. Oviparous. Distributed throughout mainland Australia, Papua New Guinea and adjacent islands.

Family: *Scincidae*

Carlia tetradactyla (O'Shaughnessy)

1879 *Mocoa tetradactyla* O'Shaughnessy
*Ann.Mag.Nat.Hist.*Ser.**V**–4:300

Localities A.C.T. Kambah, Tharwa
N.S.W. Tumut, Gunning, (Jugiong)
Vic. nil

Diagnosis This species is robustly built with well-developed limbs. It has four fingers and five toes. The lower eyelid contains a palpebral disc which is larger than the ear opening. There are six supraciliaries. The suture between the rostral and the frontonasal is about equal in width to the frontal. The prefrontals are separated. The frontoparietal, the largest head shield, is single. The interparietal is distinct, but considerably reduced.The supralabials number seven; the anterior four have an indentation on their upper margins which forms a nasal groove with the nostril. The tympanum is distinct and round, and without any obvious lobules. The body scales are in 30 to 32 rows. They are feebly quadricarinate on the dorsum and smooth on the belly.

The head and body are pale brown with a greenish tinge. The four median rows of scales on the back are mottled black, giving the back a reticulated appearance. This pattern extends on to the proximal region of the tail before coalescing to form a dark longitudinal line which continues down the entire length of the tail. There are two broad dorso-lateral stripes of pale brown. The flanks are also mottled on the upper surface. The belly and throat are cream with a bluish tinge. The males display bright breeding colours; the blue on the throat becomes more intense, and two broad orange stripes appear on the flanks. The average length of adults is 120 millimetres (50 mm + 70 mm).

General biology *Carlia tetradactyla* is one of the more colourful lizards on the Southern Highlands. It is not a common

Carlia tetradactyla; Gilgandra N.S.W.

inhabitant of the region and appears to be restricted to the well-drained savannah country and dry sclerophyll forests of the lower altitudes. It becomes most abundant towards the western perimeter of the district.

It has diurnal habits and is often found sunning itself on a fallen log or scuttling through the grass. During the winter months, we have found individuals completely torpid inside rotten logs, in company with a number of other species of small reptile, notably the pygopodid *Delma inornata*.

The diet of this species consists of small arthropods, particularly termites, a species of which occurs quite commonly within rotten logs in the region.

Mating occurs during spring. Eggs number two to five, and these are usually laid in early December, inside or beneath a rotting log. Development takes about two months and the young appear in late summer.

Genus *Ctenotus* Storr 1964

Type species: *Lacerta taeniolata* Shaw 1790

Small to moderately large terrestrial skinks, with strong pentadactyl limbs. Auditory meatus conspicuous, with two to five anterior lobules. Supranasals and postnasals absent. Lower eyelid movable and scaly. Frontoparietals paired. Distinguished by patterns of longitudinal stripes or row of ocelli. Oviparous. Distributed throughout continental Australia and southern New Guinea.

Key to species

1   Top and sides of body with a series of longi-
    tudinal stripes and spots..(2)
    Top and sides of body with alternate series
    of longitudinal stripes, never with spots...........................*taeniolatus*

2   Dorsum with spots; prefrontals usually separ-
    ated or just in contact ....*uber orientalis*
    Dorsum never with spots; prefrontal in broad
    contact .... *robustus*

Family: *Scincidae*

Ctenotus robustus Storr   **Striped Skink**

1969 *Ctenotus robustus* Storr

*J.Proc.R.Soc.West.Aust.***52**(4):100

Localities
A.C.T.	Coppins Crossing, Canberra, Murrumbidgee River
N.S.W.	Lake George, Murrumbidgee River, Goulburn, Yass, Gunning
Vic.	Dartmouth, Bright, (Yea)

Diagnosis The general body form is slender. The tail is extremely long and whip-like. The limbs are well developed and pentadactyl, overlapping when adpressed. The lower eyelid is scaly. The tympanum is distinct with three to five anterior lobules. The supraoculars number four. There are eight supralabials. The frontal is twice as long as it is broad. The frontoparietals are divided and reduced. The body scales are smooth in 24 to 34 rows around the body. The prefrontals form a median suture.

The general body colour is sandy yellow to pale brown. There is a number of stripes on the dorsum and the tail. There are two thin white dorso-lateral lines commencing on the brow and extending down the tail. The black vertebral line, which commences on the nuchals, has a thin white margin. The two dorso-lateral stripes have a suffused black margin. The upper region of the flanks is dark brown with a series of white ocellations. Below this region, the general colour is cream with a poorly-defined mid-lateral brown line between the limbs. The belly is white, however the pelvic region is often yellow in adult specimens. The average length of adults is 300 millimetres (100 mm + 200 mm).

General biology *Ctenotus robustus* is a moderate-sized skink, common throughout the lower altitudes of the Southern Highlands region. It inhabits savannah country, and is extremely common along watercourses where it is sympatric with *C. taeniolatus* and *C. uber orientalis*.

131

Ctenotus robustus; Coppins Crossing A.C.T.

It is diurnal in habit, and is very active during the summer months. If disturbed from beneath a rock or log, it darts off quickly to another hiding place; its speed and agility making it extremely difficult to catch. *C. robustus* is usually to be found wintering beneath a rock at the end of a shallow, self-excavated burrow. The tail is autotomised if the animal is restrained or suspended by it.

Ctenotus robustus is oviparous, depositing five to seven eggs in a burrow beneath a rock. Development takes approximately two months, with the young appearing in late February.

The diet of this species consists of insects and small skinks (including its own young). *C. robustus* often falls victim to large snakes such as the Brown Snake *Pseudonaja t. textilis*.

STORR, G. M. (1971) 'The genus *Ctenotus* (Lacertilia, Scincidae) in South Australia.' *Rec.S.Aust.Mus.* **16**:1–15

Ctenotus taeniolatus (Shaw)   **Copper-tailed Skink**

1790 *Lacerta taeniolata* Shaw
               White — *J.Voy.N.S.W.*: p. 245 Pl.32, Fig. 1

Localities   A.C.T.   Black Mountain, Cotter River, Mt Majura

             N.S.W.   Murrumbidgee River and lower tributaries, Braidwood, Tarago, Yass, Goulburn, Lake George

             Vic.   (Buchan), Beechworth, Tintaldra, Dartmouth, Tongio

Diagnosis The limbs are well developed and pentadactyl, overlapping when adpressed. The general body form is slender. The tail is proportionately very long. The tympanum is distinct, with four anterior lobules. The frontal is twice as long as it is broad. The frontoparietals are divided and reduced. The interparietal is also very much reduced. There are seven supralabials, five and six being subocular. There are 24 to 26 smooth scales around the body.

The head and body are jet black with a series of white longitudinal stripes. There are two light brown dorsal stripes which gradually widen towards the tail. At the base of the tail, which in juveniles is bright orange fading to light brown in adults, the black coloration becomes reduced to just a few very narrow stripes. These lines become indistinct at the distal region of the tail. The ventral surface is white. The average length of adults is 190 millimetres (60 mm + 130 mm).

General biology *Ctenotus taeniolatus* is an extremely agile skink of moderate proportions. It is diurnal and widespread throughout the warmer areas of the Southern Highlands, being most abundant in the dry sclerophyll forests, where it may often be found on sunny days scuttling through the leaf litter. It is extremely common in the sandstone country immediately north of Marulan in N.S.W., where it may be found under slabs of sandstone, often curled up in a self-excavated depression.

Ctenotus taeniolatus is an oviparous species, depositing five eggs at the end of a burrow which is excavated beneath a rock. After oviposition, the burrow is loosely sealed up. Embryonic development takes about eight weeks and the young emerge in February.

The diet consists of insects, although large specimens often prey upon smaller species of skink.

Ctenotus taeniolatus; Beecroft Peninsula N.S.W.

Ctenotus uber orientalis[1] Storr

1971 *Ctenotus uber orientalis* Storr   *Rec.S.Aust.Mus.* **16**:8

Localities   A.C.T.   Canberra
                N.S.W.   Bunyan
                Vic.   Bright

Diagnosis   *Ctenotus uber orientalis* is a moderate-sized skink with well developed, pentadactyl limbs. Body scales are smooth and in 30 rows at the mid-body. Nasals and prefrontals are separate. The elongated frontal, which is twice as long as it is broad, is bordered posteriorly by two reduced frontoparietals. Supraoculars number four and supralabials seven. Subdigital lamellae each have a brown callous or obtuse keel.

Dorsally the colour pattern consists of a series of longitudinal stripes and spots in the following order: a narrow black vertebral line, one scale wide, extends from the nuchal region to the tail, becoming indistinct on the distal portion; the vertebral line is bordered on both sides by thin white stripes; rows of small white ocelli are enclosed in two dark mid-lateral stripes; two distinct dorso-lateral white lines extend from the eyebrows to the base of the tail. On the flanks, there is a broad dark upper lateral zone enclosing a series of small white spots; an irregular, pale mid-lateral line, formed by the coalition of white blotches, is bordered ventrally by an indistinct dark line. Fore and hindlimbs are prominently striped. The ventral surfaces are immaculate white. Average length of adults is 268 millimetres (57 mm + 211 mm).

General biology   *Ctenotus u. orientalis* appears to have a restricted and discontinuous distribution in the highlands of south-

---

[1] Although we believe that the use of sub-species should generally be avoided, we have chosen to accord this status to *C. uber*. The eastern race of this lizard differs sufficiently from the nominate race to warrant citation as the sub-species *orientalis*. Further taxonomic work may show *C. uber orientalis* to be a distinct species.

eastern Australia, where it is confined to the lower altitudes. It inhabits grasslands and savannah woodlands that have been partly cleared, and is associated with granite rock outcrops.

Although this skink was known to be in the region, all records were, until recently, from the early 1900s. Several specimens registered in the Australian Museum were collected by S. Copland in 1943 from 13 kilometres north of Cooma. The National Museum of Victoria has in its collection one specimen of *C. uber* from Bright, collected in 1905 by W. H. Davey. In 1974, we recorded one specimen (now lodged in the Australian National Wildlife Collection at CSIRO, Canberra) from MacGregor, a western suburb of Canberra. Outcropping granite rock is a common and widespread feature of the Monaro Plains and associated tablelands; it is therefore probable that future collecting will show this species to have a much wider distribution in the region than it is now accorded.

The absence of a distinct pale mid-lateral stripe, which is a major character used to separate this species from *C. robustus*, does not apply when used for eastern populations of *C. u. orientalis* and the identification should, therefore, only be made with caution. *C. u. orientalis* has diurnal habits and is sympatric with *C. robustus*, the species with which it is most likely to be confused.

In common with *Ctenotus* species, *C. u. orientalis* is capable of fast turns of speed during warm periods of the year, but during the winter months it may be found under rocks in a torpid state.

It is oviparous, although nothing is known of the number of eggs laid and incubation period.

Ctenotus uber orientalis; Canberra A.C.T.

Genus *Egernia* Gray 1839

Type species: *Scincus whitii* Lacépède 1804

Palatine bones do not meet on the median line of the palate. Eyelids well developed and scaly. Tympanum distinct and sunken. Parietal shields not in contact behind interparietal. Supranasals absent. Nostrils pierced in the nasal, which may be divided by a vertical groove. Limbs pentadactyl and well developed; digits cylindrical or compressed, with transverse subdigital lamellae. Viviparous. Distributed throughout Australia, Tasmania and Papua New Guinea.

Key to species

1 Dorsal and caudal scales smooth or keeled,
   but never strongly spinose .... (2)
   Dorsal and especially the caudal scales
   strongly spinose .... *cunninghami*

2 Dorsal body scales strongly keeled .... *saxatilis*
   Dorsal body scales smooth .... *whitii*

Family: *Scincidae*

Egernia cunninghami (Gray)   **Cunningham's Skink**

1832 *Tiliqua cunninghami* Gray   *Proc.R.Zool.Soc.***40**

Localities A.C.T. Canberra, Kambah, Coppins Crossing, Uriarra Crossing

N.S.W. Queanbeyan, Gundaroo, Yass, Collector, Goulburn, Gunning, Braidwood, Lake George, Tumbarumba, Kosciusko National Park

Vic. Suggan Buggan, Wulgulmerang, Gelantipy, Omeo, Strathbogie Ranges, (Yea), Dartmouth

Diagnosis *Egernia cunninghami* is a large robustly-built species. The chin shields are large and prominent. The frontal is small and narrow, being slightly larger than each of the frontoparietals. The nuchals and anterior body scales are multicarinate. The tympanum is large, and bears four anterior lobules. The dorsal and lateral body scales are strongly unicarinate, each keel terminating in a spine. The dorsal and lateral caudal scales are strongly spinose. All ventral scales are smooth. The four limbs are well developed and pentadactyl; each digit is equipped with a claw.

The dorsal coloration is either light or dark brown with variegations. There is a scattering of white scales, which become more numerous on the flanks. The ventral surface is cream or pink with black variegations. Some of the head plates, particularly the temporals and the posterior supralabials, have white spots. The average length of adults is 430 millimetres (230 mm + 200 mm).

General biology This species is perhaps the most ubiquitous large skink in low altitude regions of the Southern Highlands.

It is gregarious with diurnal habits; it occurs often in large groups amongst granite outcrops, where it is frequently seen basking in the sun. If disturbed, it retreats rapidly into a nearby rock crevice from which it can be extremely difficult to dislodge because of the spinose nature of its scales. This wedging effect is increased by the animal inflating its lungs to swell its body in order to fit more tightly between the walls of the fissure. In timbered country, it is often encountered sunning itself on a hollow log or a tree stump, into which it will retreat to seek refuge when approached.

The diet of *E. cunninghami* includes a wide range of invertebrates, including land molluscs. This diet is supplemented by vegetable matter.

The young are free-born and usually appear at the end of summer. The coloration of the juveniles appears strikingly different from that of the adult. The basic colour is a very dark steely blue, which accentuates the scattered creamy gold scales on the back and flanks. The litter size may number from three to eight.

Egernia cunninghami; Canberra A.C.T.

Egernia saxatilis intermedia[1] Cogger   **Black Rock Skink**

1960 *Egernia saxatilis intermedia* Cogger

*Rec.Aust.Mus.***25**(5):96

Localities	A.C.T.	Gibraltar Falls, Tidbinbilla Nature Reserve, Bulls Head
	N.S.W.	Coree Flats, Braidwood, Tumbarumba, Tinderry Ranges
	Vic.	Suggan Buggan, Gelantipy, Mt Wills, Strathbogie Ranges, Shelley, Tolmie, (Yea), Dartmouth, Benambra, Murrindal

Diagnosis   This species is a moderate-sized skink with well-developed pentadactyl limbs; each digit is equipped with a strong claw. The forelimbs and hindlimbs overlap when adpressed. The tympanum is distinct with four rugose anterior lobules. The body scales are in 28 to 35 rows, 32 being the average number. The dorsal and lateral scales are predominantly bicarinate or tricarinate, although there is a scattering of quadricarinate scales. The caudal scales and nuchals are multicarinate. The supralabials number eight; the supraoculars number four, the first two being in contact with the frontal. The prefrontals make contact with one another. The frontoparietals are divided and in contact, forming a median suture immediately behind the elongate frontal.

The head, tail and dorsum are chocolate brown with a number of small black blotches which form a series of discontinuous longitudinal lines on the back. The flanks are dark brown or black with a number of light brown spots. The melanic colouring is most intense in the region of the shoulders. The infralabials, ear lobules and the supralabials

---

[1] The nominate species '*saxatilis*' is confined to the Warrambungle Mountains of New South Wales; in the remainder of its range (including the Southern Highlands) this species is accorded the sub-specific name '*intermedia*'.

are cream. The belly is salmon pink. The average length of adults is 230 millimetres (110 mm + 120 mm).

General biology This species is a common inhabitant of the rocky montane country of the Southern Highlands region. The habits of *E. intermedia* are similar to those of *E. cunninghami*. The latter species, which inhabits the lowlands country, is replaced by this species in the mountainous country. The absence of sympatry alleviates the possibility of competition by two species with similar habits, occupying similar ecological niches.

Egernia intermedia uses the many rock crevices common in the granite country that it inhabits, and may frequently be encountered as it sunbasks on a rock ledge adjacent to a crevice. Once the lizard has retreated into a crevice, it is extremely difficult to extricate because of the rugose nature of the scales and the lizard's tendency to inflate itself and become wedged against the sides of the crevice.

The Black Rock Skink retains the young within her body until the embryos are fully developed, before giving birth to two live young. The diet of this species includes a wide variety of invertebrate animals including insects.

COGGER, H. G. (1960) 'The ecology, morphology, distribution and speciation of a new species and subspecies of the genus *Egernia* (Lacertilia, Scincidae).' *Rec.Aust.Mus.* **25**:95–105

Egernia saxatilis intermedia; Timbillica State Forest via Eden N.S.W.
Photo courtesy of J. C. Wombey

Family: *Scincidae*

Egernia whitii (Lacépède)　　　　　**White's Skink**

1804 *Scincus whitii* Lacépède

Ann.Mus.d'Hist.Nat.Paris 4:192

Localities　A.C.T.　　Bulls Head, Mt Coree, Mt Franklin, Mt Ginini

N.S.W.　　Captains Flat, Snowy Range, Cooma, Mt Kosciusko, Bombala, Tom Groggin, Delegate, Adaminaby, Berridale, Jindabyne

Vic.　　Native Dog Plain, Buchan, Mt Cobberas, Mt Buffalo National Park, Tintaldra, Dartmouth, Mt Murphy, Strathbogie Ranges, Shelley

Diagnosis　This moderate-sized skink is robustly built with well-developed pentadactyl limbs. There are eight supralabials, numbers six and seven being subocular. The tympanum is distinct with three to five anterior lobules, decreasing in size downwards. The prefrontals make contact with one another. The frontoparietals are paired. The interparietal and the frontal are about the same size. There are four supraoculars. The limbs overlap when adpressed. The lower eyelid is scaly and opaque. The body scales are smooth in 32 to 40 rows.

The head, tail and body are light brown. There are two black para-vertebral lines, each containing a number of small white ocellations. The shoulder, tympanum and lip are white; each white mark has a black margin. The belly is creamy-white. Some individuals may be uniform grey or dark brown and may be devoid of any of the above described markings, except for the white colouring in the region of the shoulders and ear openings. The average length of adults is 200 millimetres (75 mm + 125 mm).

General biology　*Egernia whitii* appears to be restricted to the montane regions of the Southern Highlands, where it occurs amongst fallen timber and low vegetation.

It is heliothermic, as are most species of reptile that inhabit the mountainous regions in the district, and may often be encountered basking in the sun. This is an extremely timid lizard which requires very little provocation to cause it to retreat to the safety of one of the many fissures that occur in the fallen timber. Each digit is well equipped with a strong claw, which enables this species to be as adept on fallen timber as it is on the ground. On being handled this species displays an aggressive nature, often biting its captor. The bite, however, is mild as the teeth are small and the jaws are not strong. If handled roughly, this species is also able to shed its tail voluntarily.

The diet consists of most of the indigenous invertebrates, although large individuals are known to feed upon smaller species of skink that inhabit the area. *E. whitii* produces two or three live young each summer.

STORR, G. M. (1968) 'Revision of the *Egernia whitei* species-group (Lacertilia, Scincidae).' *J.Proc.R.Soc.West.Aust.***51**:51–62

Egernia whitii; Kangaroo Valley N.S.W.
Photo courtesy of I. Morris

Genus *Hemiergis* Wagler 1830

Type species: *Tridactylus decresiensis* Cuvier 1829

Slender body covered with smooth scales. Vermiform habit. Transparent palpebral disc contained in lower eyelid. Prefrontals absent or present and prominent. Tympanum either covered with scales or minutely punctiform. Limbs reduced, failing to overlap when adpressed; each with five digits or less. Restricted to the temperate regions of Australia and adjacent islands; absent from arid zones.

Inclusion of *maccoyi* into this genus follows Cogger (1975) and is somewhat arbitrary pending placement into a more appropriate genus than the Polynesian *Anotis* Bavay.

Key to species

Fingers and toes, three... *decresiensis*
Fingers and toes, five .... *maccoyi*

Hemiergis decresiensis (Cuvier) **Three-toed Skink**

1829 *Tridactylus decresiensis* Cuvier
(ex. Péron MS) *Règne Anim.*2nd edn.vol.2:64

Localities

A.C.T.	Mt Ainslie, Honeysuckle Creek
N.S.W.	Michelago, Orroral Valley, Braidwood, Bungendore, Collector, Tarago, Goulburn, Gundaroo, Cooma, Adaminaby, Mongarlowe, Tumut, Talbingo, Mt Kosciusko, Cullerin, Wee Jasper, Tooma, Tumbarumba
Vic.	Honeysuckle Track, Gelantipy, (Buchan), Dartmouth, Shelley, Wulgulmerang, Strathbogie Ranges, Tolmie, (Yea)

Diagnosis The body is long and slender with reduced limbs; each has three fingers or toes. There is a narrow suture between the rostral and frontonasal separating the large nasals. There are no supranasals. The frontoparietals are paired, being subequal in size with the interparietal. The parietals are the largest head shields. There is a large palpebral disc contained in the lower eyelid. The tympanum is represented by a slight depression. The body scales are smooth in 22 rows. There are seven lamellae beneath the mid-toe. There are four enlarged preanal scales, the median two being the largest.

The coloration is very variable; the basic pattern is grey or chocolate brown with a series of black longitudinal lines extending the entire length of the body and tail. The head shields are brown with black mottlings. The scales on the belly and the ventral surface of the tail are orange with a black margin. There is a reticulated pattern on the throat. The average length of adults is 145 millimetres (65 mm + 80 mm).

General biology This small species has a wide distribution on the Southern Highlands, not being confined to any one particular habitat type. It ranges from altitudes as low as 500

147

metres to the montane regions at altitudes exceeding 1800
metres.

It is an entirely terrestrial species and is unable to climb
even the smallest shrubs as do many of the smaller skinks.
It is most frequently to be found under fallen timber or rocks
where it feeds on small soft-bodied arthropods. When first
uncovered from beneath a log, this lizard remains motionless;
after several seconds, it moves off rapidly with sinuous movements through the grass and rubble under the log, and is
extremely difficult to check as it is a capable burrower.

This species, as do all members of the genus, gives birth
to live young, which may number from two to five. The young
usually appear in late summer.

COPLAND, S. J. (1945) 'Geographic variation in the lizard *Hemiergis decresiensis* (Fitzinger).' *Proc.Linn.Soc.N.S.W.* **70**:62–92

Hemiergis decresiensis; Hall A.C.T.

Hemiergis maccoyi (Lucas and Frost)

1894 *Siaphos maccoyi* Lucas and Frost
Proc.R.Soc.Victoria,(N.Ser)**6**(24):92 Pl.2,Fig.2

Localities

A.C.T.	Coree Flats, Brindabella Ranges, Piccadilly Circus
N.S.W.	Talbingo, Mongarlowe, Monga, Bondo State Forest, Mt Kosciusko, Bombala, Batlow
Vic.	Walhalla, Loch Valley, Wulgulmerang, Noojee, Goongerah, Powelltown, Neerim, Aberfeldy, Dartmouth, Mt Hotham, Mt St Leonard, Bright, Shelley, Tolmie, Bonang, Gelantipy, Mt Buffalo National Park

Diagnosis The limbs, although reduced, are pentadactyl. The general form is elongate. There is a large palpebral disc contained in the lower eyelid. There are 22 smooth scales around the body. The tympanum is minute and punctiform. The suture between the rostral and the frontonasal is almost equal in width to the frontal. The frontoparietals are paired. The snout is sharp.

The dorsal surface is dark russet with small cream ocellations randomly scattered over the entire surface, including the tail. There are two narrow dorso-lateral stripes that originate from a point above the tympanum and extend along the entire length of the body, gradually becoming indistinct on the tail. The belly is cream. There is no sexual dimorphic colouring. The average length of adults is 105 millimetres (45 mm + 60 mm).

General biology This small skink appears to be restricted to the wet gullies and swamps in the high mountains that are quite extensive throughout the Southern Highlands.

Hemiergis maccoyi is nocturnal in habit. During the day it may be found either under water-soaked logs or in amongst

149

the tangled mass of sphagnum moss. It is sympatric in these habitats with the small frog *Pseudophryne corroboree*. *H. maccoyi* is able to remain active at the extremely low temperatures which are a feature of its range; the degree of activity is minimal, and the species cannot be described as being particularly agile.

Very little is known of the habits of this secretive little skink, other than that its diet consists of small insects, and that it is oviparous. The female produces two to four soft-shelled eggs which are often deposited in communal sites under logs, or buried in moss where there is a high level of moisture. These sites are frequently shared with *Leiolopisma delicata*.

Hemiergis maccoyi; Brindabella Ranges A.C.T.
Photo courtesy of I. Morris

Genus *Leiolopisma* Duméril and Bibron 1839

Type species (by monotypy): *Scincus telfairi* Desjardin 1831

Lower eyelid movable, with a transparent palpebral disc. Tympanum distinct. Supranasals absent. Rostral forms a suture with the frontonasal. Frontal not broader than the supraocular region. Enlarged nuchals present. Limbs well developed and pentadactyl. Oviparous and live-bearing species. Distributed throughout Australia generally.

Leiolopisma is a polyphyletic and cosmopolitan genus. In Australia, the genus comprises egg-laying and live-bearing species, generally with a high degree of sympatry. Most authorities on the taxonomy of skinks of the sub-family Lygosominae agree that oviparous species more appropriately belong to Cope's genus *Lampropholis*, confining *Leiolopisma* to the group of live-bearing skinks.

Key to species

1 Frontoparietals fused ....(2)
  Frontoparietals paired...(6)

2 Suture between rostral and frontonasal much
  narrower than the frontal...(3)
  Suture between rostral and frontonasal about
  as broad as frontal...(5)

3 Five supraciliaries, occasionally six; dark lateral zone bordered above by a well-defined,
  narrow whitish line; if the latter is absent, the
  dorsum is immaculate, without a series of
  darker mottling, spots or stripes ....(4)
  Seven supraciliaries, occasionally six or
  eight; dark lateral zone not bordered above
  by a well-defined, narrow whitish line; dorsal
  pattern with darker spots, mottlings or
  stripes...*metallica*

4 Dorsal pattern includes a distinct pale dorso-
  lateral stripe and usually a dark vertebral
  stripe or other dorsal markings; usually a pale
  mid-lateral stripe.. *trilineata*
  Dorsum uniform or with faint longitudinally
  aligned rows of small black spots; distinct
  pale dorso-lateral and mid-lateral stripes
  absent.. *platynota*

5 An irregular dark vertebral stripe; dorsal surface usually with numerous white mottlings
  and dots; dark lateral zone bordered below
  by a well-defined narrow line lighter in colour
  than the dorsal surface.. *guichenoti*
  No dark vertebral stripe; dorsum either
  immaculate or with numerous, small dark
  flecks; distinct white spots absent; dark lateral zone not bordered below by well-defined
  narrow light-coloured stripe ....*delicata*

6 Suture between the rostral and frontonasal
  about as broad as the frontal...(7)
  Suture between the rostral and frontonasal
  much narrower than the frontal ....(8)

7   Dorsal surface dark brown; broad black lateral band present...*coventryi*

Dorsal surface russet-brown with numerous paler-coloured spots; broad black lateral zone absent ....*mustelina*

8   Dorsal surface light brown; pale-coloured longitudinal stripes present on dorsum ....*entrecasteauxii*
Form A

Dorsal surface dark brown; pale longitudinal stripes absent ....*entrecasteauxii*
Form B

Leiolopisma coventryi Rawlinson

1975 *Leiolopisma coventryi* Rawlinson
Mem.Natl.Mus.Victoria, Melbourne,**36**(2):1–16

Localities A.C.T.   Mt Ginini, Piccadilly Circus, Coree Flats, Bulls Head

           N.S.W.   Bondo (near Tumut), Micalong Swamp, Mt Kosciusko, Thredbo, Cabramurra, Island Bend, Sawpit Creek, Eucumbene, Batlow

           Vic.   Mt Cobberas, Wulgulmerang, Gelantipy, Goongerah, Strathbogie Ranges, Mt St Leonard, Dargo High Plains, Mt Baw Baw, Shelley, Native Dog Plain, Tolmie

Diagnosis *Leiolopisma coventryi* has a slender form with well-developed, pentadactyl limbs. The limbs when adpressed fail to, or just meet. There are 26 to 29 rows of smooth, feebly-striated scales in the mid-body region. The frontoparietal is divided. There are four supraoculars, the second being the largest. The frontal and interparietal are longer than they are wide. The parietals are bordered posteriorly by two large nuchals, which are then followed by a series of smaller pairs. The rostral and frontonasal are in broad contact; frontonasal and frontal are in narrow contact.

The dorsal colour is dark chocolate-brown. The ventral surface is light grey. There is a broad black lateral band originating at the snout and passing through the orbit and shoulder. There is a thin line delineating the upper margin of the lateral band. It is frequently present only in the region of the shoulders or immediately anterior to them. The lower margin of the band is indistinct. The head shields, supralabials and infralabials and gular region are marked with numerous small dark spots. The palms and subdigital lamellae are black. The average length of adults is 101 millimetres (41 mm + 60 mm).

General biology Superficially this species resembles *L. delicata* with which it is sympatric. It appears to be confined to the forested areas of the Southern Highlands, both in New South Wales and in Victoria. It has also been recorded in the Blue Mountains of New South Wales. It occurs in wet and intermediate sclerophyll forests and in sub-alpine woodlands, and is reasonably abundant during the summer months in leaf litter and other ground debris. The densest populations of *L. coventryi* seem to occur in naturally-burnt, regenerating montane forests.

This lizard overwinters in a state of torpor within rotten logs, often in groups, as do most other reptiles that inhabit the montane regions in the district. These groups often include other species of scincid lizards such as *Sphenomorphus tympanum* and *L. entrecasteauxii* Form B.

The diet of *L. coventryi* consists of small insects and other forms of small terrestrial invertebrates.

Leiolopisma coventryi produces from one to seven live young, three being the average. Mating occurs in autumn and the female retains the sperm over the winter period. Fertilisation of the eggs follows ovulation which occurs in late October or early November. The gestation period is of approximately 12 weeks; the young appear in February.

RAWLINSON, P. A. (1975) 'Two new lizard species from the genus *Leiolopisma* (Scincidae, Lygosominae) in South Eastern Australia and Tasmania.' *Mem.Natl.Mus.Victoria, Melbourne,* **36**(2):1–16

Leiolopisma coventryi; Mt St Leonard Vic.

Leiolopisma delicata (De Vis)

1888 *Mocoa delicata* De Vis   *Proc.Linn.Soc.N.S.W.***2**:820

Localities A.C.T.   Canberra, Piccadilly Circus, Honey-suckle Creek, Orroral Valley, Black Mountain, Coree Flats

            N.S.W.   Cooma, Lake Eucumbene, Braidwood, Charleyong, Mongarlowe, Bombala, Sassafras, Michelago, Shoalhaven River, Badger State Forest, Nimmitabel

            Vic.   Goongerah, Omeo, Dartmouth, Murrindal, Dargo

Diagnosis   *Leiolopisma delicata* is a small, though stoutly-built, species with well-developed pentadactyl limbs. The suture between the rostral and the frontonasal is as wide as the frontal. The supraciliaries usually number seven. The frontoparietals are fused to form a single shield, the largest of the head shields. There are 22 to 28 rows of smooth scales around the body.

     The dorsal coloration is a shiny dark metallic brown which is either immaculate, or has small black dots, or a faint dark brown vertebral stripe; the snout region may be paler. There is a dark lateral band from the nostril to the level of the vent, with a fine pale cream line along the upper edge and fading to pale brownish grey below. There may be a pale cream line or a series of small black dots just below the dark lateral band. The head is dark grey down to the upper lip; both lips are bordered with blackish dots. The belly and the lower jaw are pale brownish grey, and the tail is slightly darker. The young have the same colouring as the adults. Average length of adults is 90 millimetres (40 mm + 50 mm).

General biology   The habits of this small skink are similar to those of its close relative, *L. guichenoti*. *L. delicata* is an extremely ubiquitous lizard on the Southern Highlands, being common both in the lowlands and in the mountains, where it is found amongst the fallen timber and vegetation adjacent to highland soaks and swamps. There it is sympatric with *L. entrecasteauxii* and *L. trilineata*. In dry sclerophyll forests, it occurs in sympatry with *L. guichenoti* and *Morethia boulengeri* amongst the dead leaves and bark around the bases of some eucalypts. Its diet consists of most small soft-bodied invertebrates.

Leiolopisma delicata is an egg-laying species, depositing two to five soft-shelled eggs under a rock or log in early summer. The developmental period is about two months.

Leiolopisma delicata; Beecroft Peninsula N.S.W.

Leiolopisma entrecasteauxii (Duméril and Bibron) Form A

1839 *Lygosoma entrecasteauxii* Duméril and Bibron

*Erpét.Gén.*5:717

Localities	A.C.T.	Orroral Valley, Brindabella Ranges
	N.S.W.	Michelago, Bungendore, Captains Flat, Nimmitabel, Bredbo, Mt Kosciusko, Lake Eucumbene, Kiandra, Berridale, Jerangle, Tantangara Dam, Tinderry Ranges
	Vic.	Gelantipy, Square Flat, Mt Wombargo, Mt Baw Baw, Dargo High Plains, Wulgulmerang, Honeysuckle Track, Daveys Plain, Mt Nunyong, Tea Tree Range

Diagnosis *Leiolopisma entrecasteauxii* is a stout-bodied lizard with well-developed pentadactyl limbs. Form A superficially resembles *L. trilineata*. The tympanum is punctiform. The palpebral disc is nearly as large as the eye itself. There are four supraoculars of which the second is largest. The frontoparietals are paired and distinct. There are six to eight supralabials; four, five and six are subocular. There is a narrow suture between the rostral and the frontonasal, although this character is variable. When adpressed the limbs make contact with each other. There are 26 to 34 rows of scales around the body. The dorsal scales are feebly bistriate.

The dorsal colour is light brown, often with a greenish tinge. There is a series of cream, dark brown or black vertebral and dorso-lateral stripes; these stripes become indistinct on the tail. The belly is cream or light grey, and may also often have a tinge of green. Mature males are distinguished by two red lateral streaks which run through the shoulder, along the flanks, and through the pelvis to the proximal region of the tail; this condition is variable, and is most intense during the mating season. The average length of adults is 135 millimetres (55 mm + 80 mm).

General biology *Leiolopisma entrecasteauxii* Form A inhabits open country adjacent to wet and intermediate sclerophyll forests, both in mountainous and in low-lying country. It also occurs in heathlands and sphagnum bogs.

It is diurnal in habit, and may often be found basking in the sun or scuttling through the grass in search of food. During the winter months it may be found either in or under rotten timber, where it is sympatric with *Hemiergis decresiensis* and *L. trilineata*. This lizard is able to shed its tail voluntarily when alarmed. The mechanism of tail autotomy is one of defence, primarily against predators, which for this species are small snakes and birds.

The diet of this species consists of small insects.

Leiolopisma entrecasteauxii displays a modification of the viviparous condition. Mating occurs in mid to late summer. The female retains the sperm over autumn and winter, ovulating late in the following spring. The gestation period is about 10 to 12 weeks, and birth occurs in late summer, approximately one year after copulation.

Leiolopisma entrecasteauxii (?) Form B

Localities A.C.T. Piccadilly Circus, Bulls Head, Coree Flats, Brindabella Mountains

N.S.W. Lake Eucumbene, Bombala, Mt Kosciusko, Tantangara Dam, Long Plain

Vic. Daveys Plains, Omeo, Mt Nunyong, Mt Cobberas, Mt Baw Baw, Mt Gibbo, Mt Pinnibar

Diagnosis There is no difference in scalation between Form A and Form B; the major difference is in body pattern and coloration.

The dorsal colour is dark brown to almost black. The flanks are coloured black. Above the broad lateral band there is an almost indistinct light-coloured line, which is most prominent above the shoulders. A thin black vertebral stripe which becomes indistinct in the region of the hindlimbs may be present. There is a white line passing through the supralabials and

Leiolopisma entrecasteauxii Form B; Piccadilly Circus A.C.T.

the tympanum, extending through the shoulders and along the flanks, and becoming indistinct anterior to the hindlimbs. During the breeding season, the belly and shoulders of mature males become bright red. There is no difference in size between the two forms.

General biology This lizard is one of the most common species in the timbered montane country of the Southern Highlands region. Its habits differ markedly from those of Form A. Whereas Form A is solely terrestrial, being found predominantly in open country, Form B is most common in wet and intermediate sclerophyll forests of the mountainous regions. It does, however, occur in alpine meadows where it displays similar habits to Form A. In timbered country this form is semi-arboreal. It is likely that it is in direct competition with *Sphenomorphus tympanum* Cool Temperate Form (C.T.F.) which displays similar habits and exploits similar micro-habitats; body size would be the only factor that might tend to offset any competition between these two species.

The diet and mode of reproduction are the same as for the preceding Form. The number of young produced is also two to six.

Leiolopisma entrecasteauxii Form A; Orroral Valley A.C.T.

161

Family: *Scincidae*

Leiolopisma guichenoti (Duméril and Bibron)

1839 *Lygosoma guichenoti* Duméril and Bibron

Erpét.Gén. **5**:713

Localities A.C.T.   Canberra, Brindabella Ranges, Piccadilly Circus, Coppins Crossing, Coree Flats

          N.S.W.   Bombala, Tumut, Mt Kosciusko, Goulburn, Yass, Michelago, Braidwood, Tinderry Mountains, Tumbarumba, Numerella

          Vic.   Suggan Buggan, Wulgulmerang, Goongerah, Gelantipy, Buchan, Walhalla, Woods Point, Mt Buffalo National Park, Dartmouth, Mt Murphy, Dargo High Plains, Mt Buller, Bright, Shelley, Strathbogie Ranges

Diagnosis   *Leiolopisma guichenoti* is a slightly-built lizard with well-developed pentadactyl limbs. There are usually 25 to 34 smooth scales around the body. The supraciliaries usually number six. The suture between the rostral and the frontonasal is approximately equal in width to the frontal. The frontoparietals are fused.

This species exhibits a wide range of colours and body patterns, varying from pale grey to deep coppery grey. Paler dots may be present on the dorsum, usually in association with darker dots. There may be a dark vertebral stripe; if present, it extends down the tail in a broken manner. The head is often dark coppery brown. There is a dark lateral band from the nostril to the hindlimbs; it may also continue for some distance down the tail. Between the ear and the hindlimbs, there are upper and lower white margins to the band; the lower light line is always more prominent than the upper light line. Below this band, the flanks may be mottled. The belly is white, often with a greyish tinge. The average length of adults is 90 millimetres (35 mm + 55 mm).

General biology   This small skink is widespread throughout the Southern Highlands, being found in all types of habitat. It occurs in greatest numbers in the dry sclerophyll forests.

During the summer months, it is often seen in the forests scuttling through the leaf litter and decorticated bark that lies around the bases of eucalypts. During the winter months this lizard enters a state of torpor, overwintering under a rock or a log. Favourite winter sites are outcrops of rock scree where some of the fissures extend for some depth below the surface and the low temperatures are not so severe.

The diet of this species consists of small invertebrates which occur in large numbers in the microhabitats that this lizard frequents.

Leiolopisma guichenoti is an egg-laying species, producing two to five soft-shelled eggs which are deposited under a log or in amongst thick layers of humus. Development takes approximately two months. The young, which measure about 30 millimetres at birth, appear in late summer or early autumn.

Leiolopisma guichenoti; Canberra A.C.T.

Leiolopisma metallica (O'Shaughnessy)　**Metallic Skink**

1874 *Mocoa metallica* O'Shaughnessy

*Ann.Mag.Nat.Hist.***4**:299

Localities　A.C.T.　　nil
　　　　　　　N.S.W.　　nil
　　　　　　　Vic.　　　Mt Baw Baw, Mt St Leonard, Mt Tanglefoot, Toolangi, Powelltown, Aberfeldy, Loch Valley, Noojee

Diagnosis　*Leiolopisma metallica* is a slightly-built species. The pentadactyl limbs meet or slightly overlap when adpressed. The lower eyelid is movable and contains a transparent palpebral disc. The nostril is pierced in a single nasal; there are no supranasals. The suture between the rostral and the frontonasal is narrower than the frontal. Frontoparietals are fused to form a single shield. There are four supraoculars, and five to seven supraciliaries. The parietals form a median suture behind a distinct interparietal. The ear opening is round and distinct, without anterior lobules. The mid-body scales are in 24 to 28 rows. Body scales are either smooth or faintly multicarinate.

The dorsal coloration of *L. metallica* is highly variable, ranging from immaculate brown to a pattern of light and dark brown flecks, sometimes aligned into a series of three dark longitudinal stripes. A dark brown upper lateral zone, containing a number of pale brown flecks, extends from the nostril to the base of the tail. Dorsally this zone is sometimes bordered by a pale dorso-lateral streak. The lower flanks and neck are light-coloured with a series of irregular dark brown flecks. The ventral surfaces are greenish to pale grey, often with dark mottlings on the chin, throat and tail. The average length of adults is 130 millimetres (50 mm + 80 mm).

General biology　*Leiolopisma metallica* is a small terrestrial skink with diurnal habits. The species is widely distributed throughout Tasmania and the islands of Bass Strait. On these islands and on the adjacent coastal parts of southern Victoria, *L. metallica* is commonly found in low heathlands, where it may be seen basking on some of the low foliage or foraging amongst leaf litter and ground debris. It is an uncommon reptile in country above 500 metres altitude, and is restricted to the south-western border of the Southern Highlands. In high country, *L. metallica* inhabits wet sclerophyll forests where it is found sympatric with a number of other skinks (i.e. *L. entrecasteauxii*, *L. coventryi*, *Hemiergis maccoyi* and the Cool Temperate Form of *Sphenomorphus tympanum*).

In common with these species, *L. metallica* has become adapted to life in a cold environment and produces live young. The number of young varies from one to seven with a mean just below four. Copulation occurs in autumn and the sperm is retained over winter by the female. Young are born from early to mid-February the following year. *L. metallica* is truly viviparous; the developing embryos are attached to the female by means of a placenta for the exchange of nutrients and wastes between the maternal and foetal bloodstreams.

The diet of this species consists principally of insects and other small invertebrates. In turn *L. metallica* is preyed upon by birds and larger reptiles. Escape from would-be predators is enhanced by voluntary autotomy of the tail.

Leiolopisma metallica; Mt St Leonard Vic.

Leiolopisma mustelina (O'Shaughnessy)   **Weasel Skink**

1874 *Mocoa mustelina* O'Shaughnessy

*Ann.Mag.Nat.Hist.***4**:299

Localities A.C.T.   nil
          N.S.W.   Braidwood, Mongarlowe, Brown Mountain, Bombala
          Vic.   Wulgulmerang, Buchan Caves, Goongerah, Bright, Mt Hotham, Walhalla, Dartmouth, Dargo, Tolmie, Murrindal, Mt St Leonard, Shelley

Diagnosis   Both pairs of limbs are well developed and pentadactyl. The lower eyelid contains a transparent palpebral disc. The tympanum is ovoid, and is not surrounded by any obvious lobules. There are five to eight supraciliaries and four supraoculars, the second of which is the largest. The frontoparietals are paired. The suture between the rostral and the frontonasal is equal in width to the frontal. There are 21 to 28 scales around the body. The dorsal scales are feebly quadricarinate, whilst the ventral scales are smooth.

The dorsal surface of the head and body is light russet brown with a number of light-coloured ocellations. There is a white mark with a black margin immediately posterior and ventral to the orbit. The tail, which is very long, is a lighter shade of the body colour; it has a number of dark lateral and ventral stripes which, although sometimes interrupted, extend over its entire length. The ventral surface is cream. The average length of adults is 110 millimetres (40 mm + 70 mm).

General biology   This species, which superficially resembles its northern relative *L. challengeri*, is not a common inhabitant of the Southern Highlands region. It appears to be restricted to the eastern perimeter of the region, extending eastwards to the coast where it becomes most abundant. It is diurnal by habit and may be found under ground debris in areas of high rainfall and adjacent to watercourses. In these areas it often occurs in quite large numbers.

The diet of this species consists primarily of small insects.

Leiolopisma mustelina is an egg-laying species, producing three to five soft-shelled eggs. The eggs are laid under rocks and logs or in humus. This species has communal oviposition sites which are visited annually by the same individuals.

Leiolopisma mustelina; Beecroft Peninsula N.S.W.

Family: *Scincidae*

Leiolopisma platynota (Peters)

1881 *Lygosoma (Mocoa) platynotum* Peters
Sitzungsber.Ges.Naturforsch.Freunde Berlin,p.84

Localities A.C.T.   Orroral Valley, Rock Valley, Honeysuckle Creek

N.S.W.   Wee Jasper, Bondo, Lake Eucumbene,
Captains Flat

Vic.   Shelley, Suggan Buggan, Dartmouth,
Deddick, Benambra, Murrindal

Diagnosis   This stoutly-built species of moderate proportions
has pentadactyl limbs that are relatively short. The limbs
when adpressed fail to make contact. The body scales are
smooth in 25 to 31 rows. The frontoparietals are fused, forming a single large scale. The interparietal is very much
diminished. The frontal is in contact with the frontonasal.
The supraoculars number four. There are seven supralabials.
The tympanum is obvious, with two or three diminished
anterior lobules. There are four enlarged preanal scales.

The dorsum is light grey, brown or fawn. Each scale on
the dorsum has several faint black striae. There is a broad
black lateral zone, three scales wide, in the mid-body region.
This band commences on the loreals, and passes through the
orbit and shoulder, dorsal to the tympanum. It continues
along the tail becoming indistinct in the distal region. It is
well delineated along the dorsal margin, but the condition
of the lower margin is variable, being either diffuse or sharply
delineated. The ventral surface is light grey. Mature males
during the breeding season may develop a red chin and gular
region; a red streak through the tympanum may also be present. The average length of adults is 160 millimetres (70 mm
+ 90 mm).

General biology   This medium-sized skink is quite widespread throughout the Southern Highlands and tends to
favour open country where it occurs in close association with

168

fallen logs and other ground debris. It also frequents the heathlands and sphagnum bogs on the uplands. This species extends eastwards on to the east coast, and is a common inhabitant of the sandstone country around Sydney.

Leiolopisma platynota is a completely terrestrial species unlike its close relative *L. trilineata*, which displays a semi-arboreal habit. These species are sympatric in a number of localities.

The diet of this species consists of small insects and other invertebrates. *L. platynota* is oviparous and produces two to five soft-shelled eggs, three being the average number. The eggs are deposited under a rock or log in mid-summer. Hatching occurs approximately three months later; the actual incubation time is dependent on ambient soil temperature.

Leiolopisma platynota; Sydney N.S.W.

Leiolopisma trilineata (Gray)   **Three-lined Skink**

1838 *Tiliqua trilineata* Gray   *Ann.Nat.Hist.***2**:291

Localities A.C.T.   Honeysuckle Creek, Coree Flats. Orroral Valley, Bulls Head, Piccadilly Circus

N.S.W.   Wee Jasper, Michelago, Captains Flat, Lake George, Bondo, Tantangara, Rules Point, Kiandra, Gundaroo, Tom Groggin, Kosciusko National Park

Vic.   Suggan Buggan, Mitta Mitta, (East Buchan), Mt Wombargo, Dartmouth, Mt Murphy, Bentleys Plain, Mt Buffalo National Park, Dargo, Shelley, Benambra

Diagnosis   This thickly-built species is one of the largest members of the genus. Fingers and toes number five. The tympanum is distinct. The lower eyelid, when closed, reveals a prominent palpebral disc through which the pupil is visible. There are six supraciliaries. There is a narrow suture between the rostral and the frontonasal. The frontoparietal is a single large plate. The interparietal is reduced, being much smaller than the equivalent scale on *L. entrecasteauxii*.

The ventral surface is creamy white. The dorsum has the following arrangement of lines: a thin black vertebral stripe extending from the neck to the pelvis; then a brown line approximately two scales wide; then two more thin lines, black and white respectively, each half a scale in width; then a broad black dorso-lateral stripe, three scales in width; and below these, a thin white stripe one scale wide, and a thin black line which gradually fades into the cream ventrals. During the breeding season, the throat of the male becomes bright red. The ventral scales are smooth, but the dorsal scales are feebly tristriate and arranged in 26 or 27 rows. The average length of adults is 160 millimetres (65 mm + 95 mm).

Leiolopisma trilineata; Tidbinbilla Nature Reserve
Photo courtesy of I. Morris

General biology This species appears to be confined to the mountainous regions in the Southern Highlands, where it is found in both wooded and cleared areas.

It is diurnal in habit, being extremely active during the summer months when it is often seen basking on fallen timber or rocks. It is an agile lizard and is particularly difficult to catch on warm days. It is easily caught during the colder months, simply by searching beneath rocks and logs where it overwinters in a state of torpor. Its diet consists of insects.

Leiolopisma trilineata is an egg-laying species, and produces five to nine soft-shelled eggs, which are usually deposited under a rock or log.

Genus *Lerista* Bell 1833

Type species (by monotypy): *Lerista lineata* Bell 1833

A large group of small smooth-scaled skinks. Supranasals absent; enlarged nasals usually in contact; prefrontals either absent or, if present, small and widely separated. Lower eyelid has a transparent disc, and is either movable or totally fused above to form a permanent spectacle. Limbs vary from well developed and pentadactyl to the total absence of forelimbs, ranging through tetra-, tri-, di-, and monodactyl limbs. Body form elongate. Forelimbs (if present) will not contact the hindlimbs if adpressed. Habits cryptic or fossorial. Distributed widely throughout mainland Australia; one species in Tasmania and adjacent Bass Strait islands.

Lerista bougainvillii (Gray)   **Bougainville's Skink**

1839 *Riopa bougainvillii* Gray   *Ann.Nat.Hist.***2**:336

Localities   A.C.T.   nil

N.S.W.   nil

Vic.   Bright, Strathbogie Ranges, Tolmie, (Yea), Beechworth

Diagnosis   The general body form of *L. bougainvillii* is slender and elongate. Forelimbs and hindlimbs are short and pentadactyl. The lower eyelid is movable and contains an undivided transparent disc. Nostrils are pierced in two large, contacting nasals. The large frontonasal is broadly in contact with the rostral. Prefrontals are small and widely separated. Paired frontoparietals and interparietal are distinct. There are four supraoculars. The two preanal scales are enlarged. Ear opening is punctiform. Around the mid-body region are 22 to 24 rows of smooth scales. The tail is thick and slightly longer than the snout-vent length.

The dorsal surface is grey to grey-brown in colour; there is either no dorsal pattern, or each scale may contain a dark brown dot which together form a series of darker longitudinal lines on the back. A well-defined, broad black dorso-lateral line passes from the nostril through the eye and shoulder to the base of the tail, where it breaks up into dark lateral variegations on the tail. The lower flanks are flecked black and white. The ventral surfaces are creamy white, and the tail is generally reddish brown. The average length of adults is 150 millimetres (70 mm + 80 mm).

General biology   *Lerista bougainvillii* is the only local member of a group of cryptozoic, fossorial lizards that typically inhabit dry areas. This species is widely distributed throughout central and western districts of Victoria and adjacent parts of south-eastern South Australia. The northern limit of the range of distribution is in the country around the Warrambungle

Mountains in New South Wales, with a discontinuous distribution along the western slopes. It also occurs in Tasmania and islands of the Bass Strait. *Lerista bougainvillii* is distributed patchily throughout the Victorian highlands, being confined to low-lying areas of low annual rainfall. It is a marginal species in country above 500 metres altitude, and occurs infrequently in dry sclerophyll forests in the Snowy and Tambo River valleys and in similar areas.

In regions where it does occur, *L. bougainvillii* may be encountered in loose soil and debris under shallowly-buried exfoliating granite and logs in sympatry with *Leiolopisma guichenoti*, *Morethia boulengeri*, *Hemiergis decresiensis* and *Diplodactylus vittatus*. The diet of this species consists of small invertebrates that exploit similar microhabitats.

On mainland Australia *L. bougainvillii* is oviparous, and produces a clutch of two to four soft-shelled eggs deposited under rocks or logs. Insular populations of *L. bougainvillii* and those in Tasmania are live-bearing lizards; eggs are retained in the oviduct and hatching is internal. Mating takes place in autumn and the female retains the sperm over the winter months. The young are born in mid to late February.

Lerista bougainvillii; Yea Vic.

Genus *Menetia* Gray 1845

Type species (by monotypy): *Menetia greyii* Gray 1845

Small skink, head subquadrate. Rostral rounded; nostril lateral, pierced in an oblong nasal; supranasal absent; frontoparietals fused to form a single large rhomboidal shield. Eyelid fused and immovable. Tympanum diminutive, and may be distinct or covered with scales. Body elongate, fusiform and subcylindrical. Body scales smooth. Limbs four and short, failing to overlap when adpressed; four fingers and five toes. Tail cylindrical and tapering. Distributed throughout mainland Australia and regions of low rainfall.

Menetia greyii Gray

1845 *Menetia greyii* Gray   *Cat.Liz.Brit.Mus.* p.66

Localities   A.C.T.   Mt Majura, Mt Ainslie,
                     Coppins Crossing
           N.S.W.   nil
           Vic.   nil

Diagnosis   This species is one of the smallest Australian lizards. Its general form is slender with well-developed limbs. There are four fingers and five toes. There is no movable eyelid. The eye is not completely surrounded by granules. The pupil is round. The tympanum is anteriorly directed and partially hidden by an overlapping anterior scale. There is a single frontoparietal. The interparietal is distinct, being only slightly smaller than the frontal.

Dorsally this species is coloured metallic brown with fine black flecks scattered randomly over the entire surface. There is a black dorso-lateral line extending from the nostrils, through the eyes, and along the body to the distal region of the tail. Below this line there is a white line passing through the supralabials, the tympanic depression and the shoulder, becoming indistinct at the hindlimbs. The ventral surface is creamy white. The average length of adult females is 80 millimetres (35 mm + 45 mm), the males are somewhat smaller about 60 to 65 millimetres (28 mm + 32 mm to 30 mm + 35 mm).

General biology   Superficially this species resembles *Morethia boulengeri*. There is little known about the general biology of this lizard. Until late 1975, only three specimens had been collected from the Southern Highlands region, then a further eight specimens were collected by pit-fall trapping on Mt Ainslie in Canberra. It is possible that further field work will reveal a wider distribution and a more common occurrence for *M. greyii*. All three localities listed above are typical of

Menetia greyii
Photo courtesy of H. G. Cogger

the large tracts of savannah woodland and dry sclerophyll forest which characterise the low-altitude country of the tablelands.

Available information suggests that this species is seasonally abundant and diurnally active. The diet consists of small insects in the main.

The species is oviparous. Four gravid females were represented in the most recent collection, and each layed three eggs in late November which hatched in eight to ten weeks.

Genus *Morethia* Gray 1845

Type species (by monotypy): *Morethia anomalus* Gray

Small pentadactyl skinks with the lower eyelid immovable and transparent. Frontoparietals and interparietal normally fused into a single rhomboidal shield; supranasal and postnasal present (except in *M. lineoocellata* where they are often fused to the nasal). Tympanum distinct and anteriorly denticulated. Body scales smooth. Preanal scales larger than the surrounding body scales. Terrestrial and diurnal. Distributed throughout mainland Australia; except for wet tropics and alpine areas.

Morethia boulengeri (Ogilby)

1890 *Ablepharus boulengeri* Ogilby   *Rec.Aust.Mus.***1**:10

Localities A.C.T.   Coppins Crossing, Mt Ainslie, Black Mountain, Gungahlin
                 N.S.W.   Yass, Goulburn, Gundaroo
                 Vic.   Bright

Diagnosis   The limbs are well developed; fingers and toes number five. The lower eyelid is immovable and transparent. There are six supraciliaries. The mid-body scales are in 27 to 32 rows and smooth. The tympanum is distinct with two to four anterior lobules. Supranasals are always present, however they are much reduced. Postnasals are also present, but these may be fused to the supranasals.

The dorsal colour is uniform grey or light brown. There is a broad black dorso-lateral stripe commencing immediately posterior to the orbit, and extending as far as the vent; beneath this there is a thin white line passing through the supralabials and the tympanum, and ending at the hindlimbs; ventral to this is a narrow black line. This arrangement of lines gives the flanks a striped appearance. The ventral surface is white. The tail is bright red in the juvenile stages, but this colour fades with age. The adult tail colour is pale brown with perhaps a hint of red (sometimes coppery-coloured). Adult males often display a bright red throat which may be present all the year but becomes more intensely coloured during the breeding season. The average length of adults is 90 millimetres (35 mm + 55 mm).

General biology   This small, very active skink inhabits the savannah woodlands and dry sclerophyll forests which typify most of the lower altitudes of the Southern Highlands. During the warmer months of the year, individuals may be found, often in large numbers, scuttling through the grass or leaf litter, where it occurs sympatrically with a number of other species of small skink.

179

The diet of this species consists of small insects and spiders. Examination of the stomach contents of several specimens revealed that termites and ants form the bulk of the food taken.

Morethia boulengeri is oviparous, producing two to three soft-shelled eggs, which are usually deposited under an object on the ground. It is not unusual to uncover what appear to be communal egg sites, where it is possible to find numerous dried remains of earlier egg clusters. This suggests that these sites are used in successive seasons.

SMYTH, M. (1972) '*Morethia* (Lacertilia, Scincidae) in South Australia.' *Rec.S.Aust.Mus.* **16**:1–14

STORR, G. M. (1972) 'The genus *Morethia* (Lacertilia, Scincidae) in Western Australia.' *J.Proc.R.Soc.West.Aust.* **55**:73–79

Morethia boulengeri; Mt Ainslie A.C.T.

Genus *Pseudemoia* Fuhn 1967

Type species: *Lygosoma (Emoa) spenceri* Lucas and Frost 1894

Small to moderate-sized skinks. Head and body dorso-ventrally compressed. Lower eyelid movable with a well-developed, transparent palpebral disc. External ear opening prominent, with two or four enlarged anterior lobules. A pair of supranasals, separated by the frontonasal. Interparietals always separate. Parietals enlarged and sutured medially. Limbs well developed and pentadactyl. Dispecific genus confined to the montane country of south-eastern Australia and Pedra Blanca Rock off the south coast of Tasmania.

Pseudemoia spenceri (Lucas and Frost) Spencer's Skink

1894 *Lygosoma (Emoa) spenceri* Lucas and Frost
*Proc.R.Soc.Victoria,***6**:81–82 P1.2, Figs 1–1a

Localities A.C.T.   Bulls Head, Mt Franklin, Coree Flats, Piccadilly Circus, Mt Ginini

N.S.W.   Brown Mountain, Nimmitabel, Delegate, Thredbo, Mt Kosciusko, (Khancoban)

Vic.   Mt Delegate, Delegate River, Bendoc, Mt Baw Baw, Loch Valley, Mt St Leonard, Dargo, Mt Wellington, Benambra, Daveys Plains

Diagnosis The body is dorso-ventrally compressed. The well-developed limbs are pentadactyl, and overlapping when adpressed. The lower eyelid contains a large palpebral disc. Supranasals are present. There are six supralabials anterior to the subocular. The tympanum is distinct, with three or four anterior lobules. The frontal is in contact with the frontonasal. The divided frontoparietals form a median suture immediately anterior to a reduced interparietal. The body scales are smooth and in 40 to 45 rows.

The basic colour is jet black or dark brown, with a number of white dorsal and lateral longitudinal lines and ocellations. The two dorso-lateral stripes extend along the tail. The white ocelli are randomly disposed. The head is pale brown with dark brown markings. The ventral surface is light grey or cream. The average length of adults is 140 millimetres (60 mm + 80 mm).

General biology This small arboreal skink is confined to the high montane country of the Southern Highlands.

It is habitually diurnal, and usually may be found sunning itself on a log or scuttling about in the undergrowth in search of its diet of small insects. The dorso-ventrally compressed body of *P. spenceri* lends itself very well to its mode of life, enabling it to exploit the narrow crevices in the timber where it lives, either in search of food or for shelter.

Pseudemoia spenceri is one of a number of small skinks, inhabiting the alpine regions of the Southern Highlands, which has undergone a high degree of reproductive specialisation in order to cope with the harsh winters characteristic of that region. This species bears live young. The female retains the sperm through autumn and winter, finally ovulating late the following spring when fertilisation occurs. The period of gestation is about 10 to 12 weeks, and the female gives birth in mid-summer to its one or two young.

RAWLINSON, P. A. (1962) 'Revision of the endemic south-western Australian lizard genus *Pseudemoia* (Scincidae; Lygosominae).' *Mem.Natl.Mus.Victoria, Melbourne,* **35**:87–96 P1.5

Pseudemoia spenceri; Brindabella Ranges A.C.T.

Genus *Sphenomorphus* Fitzinger 1843

Type species: *Lygosoma melanopogon* Duméril and Bibron

Small to moderately-large diurnal species with pentadactyl limbs. Tympanum sunken, without anterior lobules. Supranasals absent and nasals undivided; parietals in contact behind the interparietal. Ovoviviparous forms. Distributed from tropical Africa through southern Asia to Australia and New Zealand.

Key to species

1 Distinct dark brown vertebral stripe absent,
   dorsal surface with irregularly scattered black
   spots...(2)
   Distinct dark brown vertebral stripe present..................... *kosciuskoi*

2 Prominent black markings on the gular
   region, belly without obvious black markings....... *tympanum* W.T.F.
   Gular region immaculate, belly covered with
   fine black spots .... *tympanum* C.T.F.

Sphenomorphus kosciuskoi (Kinghorn)

Alpine Water Skink

1932 *Hinula quoyii kosciuskoi* Kinghorn   *Rec.Aust.Mus.* **18**:359

Localities   A.C.T.   Mt Bimberi, Mt Ginini
              N.S.W.   Long Plain, Snowy Mountains, Smiggin Holes, Mt Kosciusko, Fiery Range, Jindabyne
              Vic.   Daveys Plain, Mt Cobberas

Diagnosis   *Sphenomorphus kosciuskoi* is a moderate-sized skink. The pentadactyl limbs are relatively short, failing to meet when adpressed. There are four supraoculars, two and three being the largest. The frontoparietals are paired, forming a median suture. There is a narrow suture between the rostral and the frontonasal. The prefrontals are also in contact forming a short suture. The frontal is twice as long as it is broad. The tympanum is distinct, without any obvious anterior lobules. There are two enlarged preanal scales. The body scales are smooth and arranged in 36 to 38 rows.

The dorsum and tail are grey with a greenish tinge. There is a series of black longitudinal lines, commencing on the nape and becoming indistinct at the base of the tail. The flanks are black with a series of light grey mottlings, which gradually coalesce on the lower surface. The tail is devoid of any markings except for several black flecks on the lateral surfaces. The belly is light grey with numerous small black flecks. The chin and gular region are without markings. The average length of adults is 355 millimetres (175 mm + 180 mm).

General biology   This skink, as its specific name indicates, was first recorded from Mt Kosciusko. It has a limited distribution on the Southern Highlands, and is confined to the alpine regions (altitudes above 1000 metres) where it occurs in the alpine woodlands and meadows sympatrically with the Cool Temperate Form of *S. tympanum*. Both these species occupy similar ecological niches and display similar habits,

185

being closely associated with upland creeks, wet heaths and bogs.[1] *S. kosciuskoi* is readily distinguished from *S. tympanum* by the presence of the longitudinal lines on the dorsum.

The species is semi-arboreal in habit and may often be seen during the summer months basking on fallen timber. During the winter months it may be found hibernating beneath deeply buried logs. Its diet consists of small insects.

Sphenomorphus kosciuskoi is viviparous, producing two to five young. The gestation period is 10 to 12 weeks.

[1] A slight qualification is necessary here as *S. kosciuskoi* tends to be more numerous in sphagnum bogs, while *S. tympanum* tends to be confined to more open types of alpine habitat.

Sphenomorphus kosciuskoi; Daveys Plains Vic.

Sphenomorphus tympanum (Lonnberg and Andersson)
Water Skink

1913 *Lygosoma tympanum* Lonnberg and Andersson
K.Sven.Vetenskapsakad.Handl. **52**

Two morphological forms of the species are recognised.

Cool Temperate Form (C.T.F.)

Localities A.C.T.   Bulls Head, Mt Ginini, Mt Franklin, Tidbinbilla Nature Reserve

N.S.W.   Bondo, Long Plain, Tantangara Dam, Snowy Mountains, Delegate

Vic.   Bendoc, Gelantipy, Walhalla, Suggan Buggan, Wulgulmerang, Mt Baw Baw, Loch Valley, Powelltown, Daveys Plain, Mt Buffalo National Park, Mt St Leonard, Mt Murphy

Diagnosis The pentadactyl limbs are well developed; the hindlimbs almost reach the shoulder when adpressed. The prefrontals are in contact, forming a narrow median suture. The suture between the rostral and the frontonasal is also narrow. The paired frontoparietals are also in contact with each other. There are four supraoculars. The frontal is twice as long as it is broad. The tympanum is without any obvious anterior lobules. The body scales are smooth and in 32 to 40 rows.

The head, dorsum and tail are golden brown, often with a greenish tinge. The head plates, particularly the supraoculars, are marked with black. The back may have a number of fine black flecks. The black lateral band commences on the supralabials, and becomes indistinct in the region of the hindlimbs. It is poorly delineated both on the upper and the lower margins. The randomly-displaced white flecks on the band become numerous on the lower margin. The belly and chin shields are creamy white with a number of fine black flecks. The gular region is devoid of any markings. The lateral surfaces of the tail, particularly at the base, are heavily mottled with black. The average length of adults is 405 millimetres (180 mm + 225 mm).

Sphenomorphus tympanum (C.T.F.); Brindabella Ranges A.C.T.

Warm Temperate Form (W.T.F.)

Localities

A.C.T.	Cotter River, Condor Creek, Coree Flats, Tidbinbilla Nature Reserve	
N.S.W.	Goodradigbee River, Wee Jasper, Brown Mountain, Clyde Mountain	
Vic.	Native Dog Plain, Mt Cobberas, Suggan Buggan, Goongerah, Murrindal, Tullochard Gorge, Mt Murphy, Dartmouth	

Diagnosis The bodily build of the Warm Temperate Form is similar to that of the preceding Form. The scalation is also similar except that the prefrontals of this Form are not in contact with each other and are separated by a narrow suture formed by the frontal and the frontonasal.

Although superficially the coloration appears to be similar in the two Forms, there are a number of distinctive differences. The black markings on the head plates are bolder in the Warm Temperate Form. Likewise, the black flecks on the dorsum tend to be both bolder and more numerous. The back is usually more golden than that of the Cool Temperate Form. The black region on the flanks is more sharply defined on the upper margin. There is often a definite thin light-coloured line in the region of the shoulders. The white flecks present on the flanks become more numerous towards the lower margin, finally coalescing on the belly. The gular region and the chin shields are heavily marked with black blotches. The belly is creamy white without any obvious black markings. The pelvic region and the thighs are often bright yellow. This condition is only found on adult specimens. The average length of adults of this Form is 410 millimetres (175 mm + 235 mm).

General biology Of the two recognised Forms, the Warm Temperate Form should probably be accorded independent specific status. This Form, as its name indicates, occurs in the warmer, lower altitudes of the Southern Highlands where the temperature extremes are not as severe as in the montane regions, but it does extend into the mountainous areas, where

Sphenomorphus tympanum (W.T.F.); Paddys River A.C.T.

it occurs sympatrically with the Cool Temperate Form. The latter is entirely restricted to the high country at altitudes above 1200 metres. *S. tympanum* (C.T.F.) occurs on the slopes of Mt Kosciusko, and in a number of other alpine localities where it is sympatric with *S. kosciuskoi.*

At the lower altitudes, *S. tympanum* (W.T.F.) is sympatric with the Gippsland Water Dragon *Physignathus l. howittii,* and may often be seen during the warmer months basking on a rock or vegetation beside a river or creek. Individuals which occur in the mountainous areas with the Cool Temperate Form do not appear to be quite so dependent on the presence of free water; they inhabit wet, intermediate, and dry sclerophyll forests, heathlands and sphagnum bogs.

Both Forms are semi-arboreal, occupying an ecological niche similar to that of *Pseudemoia spenceri,* the disparity in physical size probably being the only factor which would reduce competition in areas where the two species occur sympatrically.

The diet of both Forms is similar, consisting of a wide range of invertebrates and occasionally small lygosomine lizards.

There appears to be very little difference in the mode of reproduction displayed by either Form. The ova are retained until the embryos are fully developed, resulting in the birth of two to five live young. Mating occurs in spring, approximately three months before parturition.

Genus *Tiliqua* Gray 1825

Type species: *Lacerta scincoides* Shaw 1790

Tympanum distinct and deeply sunken. Supranasals absent. Pterygoids toothless; lateral teeth usually with spherical crowns. Limbs relatively short and pentadactyl. Ovoviviparous. Distributed throughout Australia and Tasmania; extending northwards to New Guinea, Moluccas and Java.

Key to species

1 Tail much shorter than snout-vent length ....(2)
  Tail nearly as long or longer than snout-vent length .... *casuarinae*

2 Anterior temporal scales more or less equal to others; not much longer than broad ....*nigrolutea*
  Anterior temporal scales much longer than others; much longer than broad ....*scincoides*

Tiliqua casuarinae (Duméril and Bibron)

Sheoak Skink

1839 *Cyclodus casuarinae* Duméril and Bibron

*Erpét.Gén.***5**:749

Localities	A.C.T.	nil
	N.S.W.	Mt Kosciusko, Smiggin Holes, Daners Gap, Kiandra
	Vic.	Mt Hotham, Mt Buffalo National Park

Diagnosis The general body form is elongate. Pentadactyl limbs are reduced in length relative to body length. The body is covered with smooth scales in 22 to 28 rows in the mid-body region. The lower eyelid is scaly and movable. Nostrils are pierced in large nasals which generally contact each other. Frontonasal is small. Prefrontals are separated, or in contact to form a median suture. There are three supraoculars, and the postnasal groove is absent. There is a pair of large preanals. The prominent external ear opening is without anterior lobules.

The colour of the dorsum varies from pale brown through olive-brown to russet, and may be immaculate or strongly patterned. The dorsal and lateral scales on some individuals have black lateral margins which give a combined impression of a series of fine black longitudinal stripes. The flanks on some individuals may be immaculate or distinctively marked with a series of black bars. These bars are formed by the transverse alignment of scales with black posterior margins. The tympanum and shoulders may also be barred, often more boldly than the flanks. The throat is either immaculate or has dark flecks. The ventral surfaces are grey to olive with several series of transversely-aligned, dark-edged scales. The orbit and subocular supralabials are often darkly marked; in some specimens, this mark may extend on to the infralabials. The tongue is dark blue. The average length of adults is 300 millimetres (130 mm + 170 mm).

General biology The Sheoak Skink is a moderate-sized, slenderly-built, terrestrial lizard with diurnal habits. It is confined to regions of high rainfall, much of which falls in winter as snow. In this country, *T. casuarinae* appears to be restricted to alpine meadows where it occurs under fallen timber and ground debris amongst tussock grass. During the colder months of the year, this lizard may be found in a torpid state; if encountered during the summer, it is able to seek cover with surprising agility, and its snake-like movement through grass is extremely difficult to arrest.

Tiliqua casuarinae is an extremely aggressive lizard. When molested, it holds its mouth agape, revealing its broad fleshy blue tongue, and often makes repeated lunges at its molestor. Its aggressiveness is directed towards other reptiles as well as to individuals of its own kind; this behaviour makes the Sheoak Skink difficult to keep in captivity with other species of reptile. Unlike the larger members of this genus, *T. casuarinae* has a long, fragile tail which is readily discarded across a number of breakage (autotomy) planes.

The Sheoak Skink is ovoviviparous. Mating occurs during the summer months and after eight to ten weeks' gestation the young are born. The litter is large, numbering six to nine. The coloration of juveniles is very variable, both geographically and within litters.

Tiliqua casuarinae; Watagan Ranges N.S.W.

Tiliqua nigrolutea (Quoy and Gaimard)   **Blotched Bluetongue**

1824 *Scincus nigrolutea* Quoy and Gaimard

*Voy.*Uranie *Zool.*p.175

Localities	A.C.T.	Brindabella Ranges
	N.S.W.	Cooma, Braidwood, Captains Flat, Hoskinstown, Nimmitabel, Tumbarumba, Lake Eucumbene
	Vic.	Suggan Buggan, Gelantipy, Bonang, Tongio West, Noojee, Mt Buffalo, Shelley, Strathbogie Ranges, Tolmie, Harrietville, Omeo, Dartmouth, Dargo High Plains

Diagnosis The general body form is robust. The limbs are pentadactyl and reduced. The head is large and triangular when viewed from above, and is distinct from the neck. The tongue is bright blue. The temporal scales are all approximately equal. There may be four to five supraciliaries and four supraoculars. The scales around the body vary between 28 and 35 in number. The dorsal and lateral scales are rugose, whilst those covering the belly are smooth. The tympanum is distinct. The short tail is thick-set and tapers abruptly to a point.

The basic body colour is glossy black on the dorsum with several rows of discontinuous blotches which extend down the tail. The belly is creamy white with black mottlings. The snout and lips are usually pale yellow. There appear to be two colour races present on the Southern Highlands: a lowlands form which has yellow blotches, and a montane variety which has pink blotches. The latter race appears to attain a greater length than the lowlands form. Average length of adults (including both forms) is 520 millimetres (330 mm + 190 mm).

General biology *Tiliqua nigrolutea* is one of the largest skinks that is to be found on the Southern Highlands. Both colour forms are restricted to the more mountainous terrain, but the smaller race (with the yellow blotches) appears to be confined to mountains and their foothills in the eastern sector of the region, whereas the larger variety is truly alpine.

The species has diurnal habits and may often be seen basking in the sun beside roadways. When approached, it scuttles off into the adjacent vegetation. It is not a very agile lizard when compared with most other skinks. The reduced limbs are of only marginal importance in locomotion.

The young are free-born, with the average litter often numbering ten or more. Birth usually occurs in mid-summer. At birth the young measure approximately 150 millimetres. Although this species is able to dismember its tail, it does so with more reluctance than do most other species of skink.

Tiliqua nigrolutea is omnivorous in its choice of food, eating insects, snails and slugs, small vertebrates and a wide range of vegetable matter.

Tiliqua nigrolutea; Brindabella Ranges A.C.T.

Family: *Scincidae*

Tiliqua scincoides scincoides (Shaw)

Common Bluetongue

1790 *Lacerta scincoides* Shaw White — *J.Voy.N.S.W.*P1.242

Localities A.C.T. Canberra, and A.C.T. generally
N.S.W. Goulburn, Lake George, (Gundagai), Cooma, Braidwood, Bungendore, Gunning, Queanbeyan, Yass, (Cootamundra), Murrumbidgee River
Vic. nil

Diagnosis The general body form is stout with reduced limbs, all of which are pentadactyl. The mid-body scales are smooth and in 34 to 38 rows. The temporals are unequal; anterior scales are larger and much longer than they are broad. The tail is relatively short and thick-set, and tapers to a point.

The general colour is grey to yellowish brown with a series of black or dark brown transverse bars. These bars extend along the entire length of the body and tail and may be either continuous or alternate. The flanks are usually a shade of brown in between the black bands. The belly is pinkish white. There are two black temporal streaks which pass through the orbit and extend laterally through the tympanum. The tongue is bright blue, the character from which the lizard derives its vernacular name. The average length of adults is 400 millimetres (260 mm + 140 mm).

General biology *Tiliqua scincoides* (the type species of the genus) is one of the most widespread and well-known lizards in the warm rural districts of the Southern Highlands. It extends into the foothills of the mountain ranges where it lives in sympatry with its close relative *T. nigrolutea*. *T. scincoides* is not as robust in build as is *T. nigrolutea*, nor does it attain the length of the latter species.

When aroused, this lizard inflates itself and faces its tormentor with its mouth agape, revealing the bright blue tongue. It expels the air periodically so producing loud hisses.

198

Although the Bluetongue has only poorly-developed conical teeth, its powerful jaws enable it to inflict a painful bite. Unfortunately this lizard often falls victim to unthinking attack by humans because of its superficial resemblance to the Death Adder *Acanthophis antarcticus antarcticus*. This resemblance is enhanced by the reduced limbs which play very little part in locomotion and are not readily discernible in long grass.

The diet of this species is similar to that of *T. nigrolutea*, consisting of snails, insects and a wide variety of vegetable matter. In captivity, individuals often display a strong liking for banana, apple and other fruits.

The young are free-born and may number as many as 20; the young are 'dropped' in late summer.

Tiliqua scincoides; Canberra A.C.T.

Genus *Trachydosaurus* Gray 1825

Type species (by monotypy): *Trachydosaurus rugosus* Gray 1825

Head large, and distinct from the neck. Tympanum distinct. Dorsal body scales large and rugose. Tail short, depressed and blunt. Limbs relatively short and pentadactyl; digits cylindrical; subdigital lamellae mostly divided. Ovoviviparous; producing one or two large young. Confined to continental Australia and islands adjacent to Perth, Western Australia. Not represented east of the Dividing Range or in Tasmania.

Family: *Scincidae*

Trachydosaurus rugosus Gray

Stump-tailed Skink
or **Shingleback**

1825 *Trachydosaurus rugosus* Gray   *Ann.Philos.*Ser.2, **10**:201

Localities   A.C.T.   Canberra
N.S.W.   Yass, Gundaroo, Collector, Sutton
Vic.   nil

Diagnosis   *Trachydosaurus rugosus* is a robust lizard with a large triangular head. The head is distinct from the neck. The reduced limbs are pentadactyl. The tail is short and bulbous with a blunt tip. The subdigital lamellae are divided. The mid-body scales are arranged in 22 to 30 rows. The scales are large and rugose on the flanks and dorsum, whilst those on the belly are smooth.

Trachydosaurus rugosus; Yass N.S.W.

So far all specimens that have been collected from the Southern Highlands have been completely melanic. The average length of adults is 380 millimetres (300 mm + 80 mm).

General biology   This lizard is one of the largest members of the family Scincidae. It is diurnal in its habits, and is characteristically to be found in the forests and woodlands of the warmer regions on the Southern Highlands. It is particularly common in partially-cleared rural areas where it may frequently be seen basking in the sun beside a disused rabbit burrow or a hollow log. *T. rugosus* is probably widely distributed throughout the warmer regions of the district, although it can not be regarded as being common in the area. West of the Southern Highlands this species is extremely abundant. It is interesting to note that individuals which occur on the Southern Highlands, where the degree of insolation is not as intense or so prolonged as it is in the arid inland, are melanic.

When captured, this lizard embarks upon an impressive defensive behaviour, inflating its lungs and expelling the air in loud hisses. During this display the mouth is held agape revealing the broad slate-blue tongue.

Trachydosaurus rugosus gives birth to live young, which typically number two. The young measure approximately 150 millimetres at birth. The diet consists of a wide range of invertebrates and vegetable matter. Because this lizard is readily tamed and its diet is commonly available, it is often kept as a pet in suburban gardens.

Monitor lizards: family *Varanidae*

Members of this family range through Africa, Asia and the western Pacific, but they have their greatest concentration and diversity of species in Australia. Varanids which occur in Australia are known colloquially as 'goannas' (a corruption of '*Iguana*'). All members of the family Varanidae belong to the single genus *Varanus*.

The general body form is elongate. The head and body are covered with small, close-fitting, juxtaposed scales. The scales of the belly and tail are larger, and tend to be arranged in annular rows. The overall nature of the skin can be described as flabby with a rough texture. The dentition is pleurodont, and the teeth are long and posteriorly recurved. The limbs are well developed, robust and pentadactyl. Each digit bears a strong claw. The pupil is round. The tongue is smooth, deeply bifid and snake-like, being retractible into a basal sheath.

Two sub-genera are recognised amongst the Australian varanids. Those in which the tail is round in section belong to the sub-genus *Odatria*. Members of this group are generally smaller than the members of the sub-genus *Varanus*. The latter are characterised by having a tail that is laterally compressed.

Goannas are diurnal lizards and vary in size from small species like *Varanus brevicauda* (approximately 150 millimetres in length) to the Komodo Dragon *V. komodoensis*, the world's largest living lizard, which attains a length of 4 metres.

Varanids are oviparous. The family includes terrestrial forms (*V. gouldii*), arboreal forms (*V. tristis*), and aquatic forms (*V. mertensi*).

Dietary preferences, which tend to be a function of size, include a wide range of animal life from insects to mammals. Many of the larger species are scavengers.

Genus *Varanus* Merrem 1820

Type species: *Lacerta varia* Shaw

Body strong. Pupil round; eyelids distinct. Tympanum distinct. Tail long, cylindrical or compressed. Teeth large with dilated bases attached to the inner surfaces of the jaw (pleurodont dentition). Tongue long and slender, smooth and deeply bifid; retractible into a sheath at the base, similar to that of snakes. Pentadactyl limbs strong. Preanal pores sometimes present. Distributed throughout Africa, southern Asia, New Guinea and continental Australia.

Varanus varius (Shaw)

Common Tree Goanna
or **Lace Monitor**

1790 *Lacerta varia* Shaw

White — *J.Voy.N.S.W.* p.246 Pl.3, Fig.2

Localities
A.C.T.	Mt Ainslie, Black Mountain, Mt Majura, Molonglo River
N.S.W.	Wee Jasper, Gundaroo, Sutton, Murrumbidgee River, Yass, Lake George, Goulburn, Braidwood
Vic.	Wulgulmerang, Omeo, Gelantipy, Noojee, Dartmouth

Diagnosis The head is large with a distinct canthus rostralis. The nostrils are oval. The dorsal surface of the head is covered with small irregularly-disposed scales. The tail is long and dorso-laterally compressed in the distal region, and keeled dorsally. All the limbs are pentadactyl, and each digit is equipped with a well-developed claw. The tongue is bifid, similar to that of snakes.

The species occurs in two distinct colour varieties between which intergrades are common. One form has broad steel-blue bands alternating with bands of cream; these bands extend the entire length of the body and tail. The other form is dark steel-grey dorsally, with a random scattering of white scales which give it a laced appearance. The belly of both forms is cream. Some specimens have yellow and blue markings around the jaws. The average length of adults is 1350 millimetres (600 mm + 750 mm).

General biology The Lace Monitor has a wide distribution on the Southern Highlands and frequents all types of country, with the exception of the alpine regions.

It is a solitary species with diurnal habits. It may often be encountered sunning itself on a log or the trunk of a tree during the hotter months of the summer. *V. varius* grows to a length of 2100 millimetres, which makes it one of Australia's largest lizards. This species, although capable of preying on animals such as birds and small mammals, shows a predilection for carrion, and may often be seen feeding on animal road victims.

Varanus varius is an oviparous species. The soft-shelled eggs are laid in large numbers inside burrows, or occasionally inside arboreal termitaria which this lizard is able to reach and excavate using its strong claws.

Even though it is not venomous, this lizard is capable of inflicting a serious bite or other wounds by lashing its tail or by clawing. Because of its habit of feeding on carrion, these wounds can become septic. Sloughing is sporadic and is never complete as in snakes and some lizards.

Varanus varius; Lake Cowal N.S.W.

Front-fanged land snakes: family *Elapidae*

This family includes all the venomous land snakes which have venom fangs which are rigidly fixed to the anterior region of the maxillary bone. The nostrils are laterally displaced. The belly scales are broad and extend to the lateral body scales. All scales are imbricate. The tail is round in cross section, tapering to a point, except in the Death Adder *Acanthophis a. antarcticus* which has a unique, laterally-compressed spine at the tip of the tail. The loreal scales are absent.

All front-fanged land snakes in Australia belong to this large family. Approximately 70 species and sub-species belonging to more than 24 genera are found in mainland Australia and Tasmania. Included in this assemblage are such toxic species as the Tiger Snake *Notechis s. scutatus* and the Taipan *Oxyuranus scutellatus*.

The Australian elapids are represented by arboreal and fossorial, diurnal and nocturnal species occupying a wide range of ecological niches.

Key to elapid genera

1 No specialised curved soft spine on tip of
tail..(2)
A curved soft spine on tip of tail ....*Acanthophis*

2 All subcaudals normally undivided ....(3)
At least some subcaudals divided...(7)

3 Frontal shield longer than broad; where
frontal is only slightly longer than broad,
lower anterior temporal shield is shorter than
frontal...(4)
Frontal shield not or scarcely longer than
broad; lower anterior temporal as long as or
longer than frontal..*Notechis*

4 Frontal less than one and one-half times as
broad as the supraocular ....(5)
Frontal more than one and one-half times as
broad as the supraocular ....(6)

5 Lateral scales adjoining ventrals not notice-
ably enlarged..*Drysdalia*
Lateral scales adjoining ventrals noticeably
enlarged ....*Austrelaps*

6 Entire dorsal and lateral surfaces of body
black or dark grey..*Cryptophis*
Dorsal and lateral surfaces of body mostly
brown or pink ....*Unechis*

7 Usually all subcaudals divided ....(8)
Usually some anterior subcaudals single, re-
mainder divided...*Pseudechis*

8 Subcaudals fewer than 35..(9)
Subcaudals 35 or more; colour pattern con-
sisting of alternate black and white bands
from head to tail...*Vermicella*

9 Nasal and preocular scales widely separated;
a bright red or orange patch or bar present
on nape...*Furina*
Nasal and preocular scales in contact; scales
in 17 rows at mid-body..*Pseudonaja*

Genus *Acanthophis* Daudin 1803

Type species: *Boa antarctica* Shaw and Nodder 1802

Head large and distinct from the neck. Body short and stout with abruptly tapering tail terminating with a spine. Fangs followed by a maximum of three maxillary teeth. Supraoculars prominent, forming two horn-like projections above the eyes. Ovoviviparous. Confined to mainland Australia and New Guinea.

Acanthophis antarcticus (Shaw)   **Death Adder**

1802 *Boa antarctica* Shaw and Nodder

*Nat.Miscell.*13:pl.*mxxxv*

Localities A.C.T.   Ginninderra
              N.S.W.   nil
              Vic.   nil

Diagnosis   Head is distinct from the neck. The stout body tapers abruptly to a thin tail. A leaf-like appendage with a terminal spine forms the distal region of the tail. The feebly-carinated mid-body scales are arranged in 21 or 24 rows. The anal is entire. The ventrals number from 118 to 123. Twenty-seven to 33 of the subcaudals are entire, while the remaining 18 to 20 on the appendage are divided. There is a single nasal in contact with the preocular. The frontal is twice as long as it is broad. The supraoculars are as broad as the frontal, appearing as horn-like projections above the eyes. The prefrontals number two. There is a distinct canthus rostralis.

Coloration is variable, being either brick red or grey with a series of darker cross-bands. The most common colour variety is grey. The belly is cream with a number of dark grey mottlings. The infralabials are barred black and white. The appendage is invariably creamy white. The average length of adults is 615 millimetres (500 mm + 115 mm).

General biology   Although only one specimen has been collected on the Southern Highlands during the preparation of this work, this species is probably quite widely distributed throughout the warmer areas of the region. It is doubtful whether it could be described as being common in any particular locality.

Acanthophis antarcticus frequents dry country where it may be found concealed in the ground litter waiting for a victim to come within striking distance. Its prey, which consists of small birds, mammals and reptiles, is lured in close by the periodic twitching of the terminal appendage of the tail. This

Acanthophis antarcticus (grey colour phase); Springbrook Q'ld.

may simulate an insect or small mammal in trouble and so attract a would-be predator.

Death Adders bear live young which may number between eight and 15. The young measure approximately 150 millimetres at birth.

Genus *Austrelaps* Worrell 1963

Type species: *Hoplocephalus superbus* Günther 1858

Large snakes, often exceeding one metre in length. Head slightly distinct from the neck. Canthus rostralis distinct; single nasal contacting the preocular; anal and subcaudals entire. Venom fang followed by three to seven solid maxillary teeth. Distributed throughout eastern Australia and Tasmania.

Acanthophis antarcticus (red colour phase); Sydney N.S.W.
Photo courtesy of I. Morris

Austrelaps superba (Günther)

Copperhead

1858 *Hoplocephalus superbus* Günther

*Cat.Colub.Sn.Brit.Mus.*3:217

Localities A.C.T. Orroral River, Gudgenby River, Coree Flats, Mt Gingera, Brindabella Ranges generally, Tidbinbilla Nature Reserve

N.S.W. Captains Flat, Hoskinstown, Tinderry Ranges, Wee Jasper, Burrinjuck Dam, Braidwood, Mongarlowe, Eucumbene, Cooma, Nimmitabel, Majors Creek, Tumbarumba

Vic. Bombala, Native Dog Plain, Gelantipy, Bonang, Wulgulmerang, Lieda, Loch Valley, Noojee, Mt Baw Baw, Mt Cobberas

Diagnosis The mid-body scales are smooth, and arranged in 15 rows. The anal is entire, and subcaudals are single, numbering 41 to 50, while ventrals number 145 to 160. There is a distinct canthus rostralis. The head is only slightly distinct from the neck. A characteristic feature of this species is the brown and white diagonal barring on the six supralabials and the lowermost temporal, giving the lip a striped appearance. The frontal is almost twice as long as it is broad and equal in width to the supraoculars. There is a single nasal in contact with the preocular. The pupil is round.

The outer edges of the anterior ventrals are coloured black. The adjacent body scales are cream coloured giving the overall impression of a dark dorso-lateral line. The colour of Copperheads is highly variable and as such is of little use diagnostically. The most common variety is dark brown or grey dorsally, with salmon-coloured lateral scales which often have black margins. The ventrals are a drab grey colour. Juvenile forms often have a black nape and vertebral stripe. These markings fade as the animal gets older. The average length of adults is 910 millimetres (760 mm + 150 mm).

214

Austrelaps superba; Tidbinbilla Nature Reserve A.C.T.
Photo courtesy of I. Morris

General biology This stoutly-built snake is typically diurnal
in its habit. It is one of the few Australian snakes that is found
above the snow-line.

Austrelaps superba is usually the last species to enter into a
state of winter torpor and the first to become active after winter. Even during the winter months on warm days it may
be encountered basking in the sun. It is usually found in
swampy country bordering rivers and lakes, where it feeds
on small reptiles and frogs, numbering amongst its prey
smaller snakes even of its own species.

It is gregarious; large numbers of individuals are often
found together under a single log or other ground debris. Copperheads are docile snakes and are reluctant to bite or even
display unless unduly molested.

A large number of live young are produced, the average
brood often being more than 24.

Genus *Cryptophis* Worrell 1961

Type species: *Hoplocephalus pallidiceps* Günther 1858

Small to moderate-sized snakes. Head is variable in width, often being considerably wider than the neck. Canthus rostralis absent. Three to five maxillary teeth follow the venom fang. Pupil round. Nasal scale in contact with preocular. Anal and subcaudals entire. Live-bearing species. Distribution confined to north-west, north and eastern Australia.

Cryptophis nigrescens (Günther)   **Small-eyed Snake**

1862 *Hoplocephalus nigrescens* Günther

Ann.Mag.Nat.Hist. **9**:131 Pl.*ix*, Fig.12

Localities A.C.T. nil

N.S.W. Braidwood, Clyde Mountains, Araluen

Vic. Wulgulmerang, (Buchan), Dartmouth, Tolmie, (Healesville), (Yea), Strathbogie Ranges, Murrindal

Diagnosis Mid-body scales are smooth and in 15 rows. Sub-caudals are all single, and number 32 to 37. The anal is entire, and ventrals number 165. A single nasal is in contact with the preocular. The frontal is slightly longer than it is broad, being much larger than the supraoculars. The pupil is round. The head is barely distinct from the neck.

Dorsally this species is uniformly jet black or grey. The ventrals may be either pink or white depending on the locality inhabited by the individual. There is a dark median line extending down the subcaudals. The average length of adults is 425 millimetres (365 mm + 60 mm).

General biology *Cryptophis nigrescens* is a small nocturnal species. It is not very common on the Southern Highlands, and is confined to the region adjacent to the coastal ranges, occurring in a wide range of habitats from wet sclerophyll forests to woodlands. Once it is east of the escarpment of the Dividing Range, this species becomes extremely abundant. It is found under rocks and other objects lying flat on the ground, while rubbish tips are a particularly good collecting site.

This snake is a gregarious species, and as many as seven specimens have been found under one piece of tin. Its diet consists primarily of small skinks of the genus *Leiolopisma*. Small-eyed Snakes seem to display a dimorphism of head shapes; the width of the head is the most variable character.

217

Cryptophis nigrescens; Tianjara Falls N.S.W.

This variation is not related to geographic trends as both forms exist together in all localities.

Although the venom apparatus of this species is small, it is capable of delivering quite a painful bite, consequently care should be exercised when handling larger specimens.

Genus *Drysdalia* Worrell 1961

Type species: *Hoplocephalus coronoides* Günther 1858

Small to moderate-sized snakes. Head slightly distinct from the neck. Canthus rostralis distinct. Pupil round. Anal and subcaudals entire. Nasal scale contacting preocular. Internasals present. Venom fang followed by three or five maxillary teeth. Distributed throughout southern Australia including Tasmania.

Drysdalia coronoides (Günther)　　　**White-lipped Snake**

1858 *Hoplocephalus coronoides* Günther

*Cat.Colub.Sn.Brit.Mus.***3**:215

Localities A.C.T.　Honeysuckle Creek, Brindabella Ranges, Bulls Head

　　　　　N.S.W.　Lake Eucumbene, Captains Flat, Tantangara Dam, Cooma, Kiandra, Mt Kosciusko

　　　　　Vic.　Loch Valley, Mt Baw Baw, Mt Cobberas, Native Dog Plain, Mt Buffalo National Park, Dartmouth, Noojee, Mt St Leonard, Omeo, Shelley, Benambra

Diagnosis The mid-body scales are feebly striated and arranged in 15 rows. The anal is entire. Subcaudals number 43 to 57; all are single. Ventrals number 134 to 156. The pupil is round. The frontal is almost three times as long as it is broad, being slightly longer than the supraoculars. The supraoculars are slightly broader than the frontal. There is a single nasal which contacts the preocular.

　The nasal, six supralabials and the lower-most temporal are white with an upper margin coloured black. This pattern extends down the body for about 25 millimetres. The lower-most body scales are striped black and white. The colour of this species is extremely variable. Dorsally it occurs in the following colours: grey, red, green or charcoal. The ventrals are not as variable in their colour as the body scales, being either red or orange in varying shades with darker peripheral regions. The most common form in the A.C.T. district is red both dorsally and ventrally. The average length of adults is 365 millimetres (290 mm + 75 mm).

General biology The White-lipped Snake is a small species, bearing a close relationship to the Master's Snake *D. mastersi*. In the Southern Highlands, it appears to be restricted to the mountainous forests, where it occurs sympatrically with the Copperhead *Austrelaps superba*.

White-lipped Snakes are often found in water-soaked areas which are common within their range. A frequent hibernation site is inside water-soaked logs. Its diet consists of small skinks which are common in the same habitat. *D. coronoides*, by virtue of the low night temperatures characteristic of its range, is necessarily diurnal in habit.

The young are retained within the female until fully developed, eventually being dropped in mid-summer. The usual number of young in a brood is three or four.

Drysdalia coronoides; Honeysuckle Creek A.C.T.

Genus *Furina* Duméril and Bibron 1853

Type species: *Calamaria diadema* Schlegel 1837

Small species. Head barely distinct from the body. Body slender and cylindrical. Tail cylindrical and attenuated. Body scales small, smooth, numerous and imbricate; anal and subcaudals divided; no canthus rostralis; nasal not in contact with the preocular. Oviparous. Distributed throughout mainland Australia; absent from the mountainous regions.

Furina diadema (Schlegel) **Red-naped Snake**

1837 *Calamaria diadema* Schlegel *Phys.Serp.***2**:32

Localities A.C.T. Canberra ?
N.S.W. Goulburn, (Marulan)
Vic. nil

Diagnosis Mid-body scales are smooth and arranged in 15 rows. Ventrals number 173 to 184. Subcaudals number 39 to 44; all are divided. The anal is divided. There is no canthus rostralis. The frontal is as long as it is broad, and is twice as broad as the supraoculars. The nasals are entire and not in contact with the preocular. The internasal is shorter than the prefrontals.

The dorsal colour is reddish-brown; each scale has a darker margin, giving the back an overall reticulated appearance. The ventrals are creamy white. Dorsally the head and neck are capped with black, which extends laterally down to the upper margin of the supralabials. The remaining part of the supralabials and the infralabials is white. Immediately behind the head plates, there is a crescent-shaped red mark from which the snake derives its common name. The average length of adults is 370 millimetres (310 mm + 60 mm).

General biology The Red-naped Snake is a small slightly-built snake. Field collecting has shown this species to be rather uncommon on the Southern Highlands, but becoming more frequent in the north-eastern part of the region, near Picton and Campbelltown.

Furina diadema is a nocturnal species which inhabits savannah woodlands and grasslands, where it may be found under logs or deeply-buried slabs of rock. An example of this type of habitat is the country near Picton in N.S.W., where slabs of mudstone lie in large numbers on the slopes of the mountain ranges. The diet consists largely of insects and small skinks.

Eggs are deposited under rocks in early summer, and juveniles may be found in mid-summer. When aroused this species adopts a menacing defensive attitude by raising its forebody off the ground in a manner similar to many of the cobras. It is extremely difficult to provoke this species into biting.

Furina diadema; Sydney N.S.W.
Photo courtesy of H. G. Cogger

Genus *Notechis* Boulenger 1896

Type species: *Naja (Hamadryas) scutata* Peters 1861

Medium to large individuals with thick-set bodies. Head is broad and distinct from the neck. Prominent canthus rostralis. Venom fangs followed by three to five small maxillary teeth. Anal is entire; subcaudals all single. Body scales in 13 to 19 rows. Ovoviviparous. Confined to temperate regions of Australia, Tasmania and off-shore islands.

Notechis scutatus scutatus (Peters) **Mainland Tiger Snake**

1861 *Naja (Hamadryas) scutata* Peters
*Monatsber.K.Preuss.Akad.Wiss.Berlin,***690**

Localities A.C.T.   Orroral Valley, Gudgenby River, Tidbinbilla Nature Reserve

N.S.W.   Lake George, Sutton, Bungendore, Collector, Tarago, Gunning, Goulburn, Hoskinstown, Lake Bathurst

Vic.   Wulgulmerang, Corryong, Benambra, Mt Baw Baw, Aberfeldy, Omeo, Dartmouth, Tea Tree Range, Mt Cobberas

Diagnosis   Mid-body scales are smooth and arranged in 19 rows. The anal is entire. The subcaudals are all single, numbering 50 to 60. The ventrals number 167 to 180. The single nasal is in contact with the preocular. The frontal is slightly longer than it is broad. The temporals are 2 + 2, the lower anterior one being very large, extending between the fifth and sixth supralabials. The head is indistinct from the neck.

Tiger snakes display a wide range of colours and patterns. Although melanic forms are found in the Southern Highlands, the more common variety is strongly banded. The body colour is olive-brown with bright yellow bands which are most prominent on the lateral surfaces; the bands become distinct on the back only when the snake becomes aroused and inflates its lung with air. The ventrals are bright yellow anteriorly, fading to grey at the posterior end of the body. The average length of adults is 1050 millimetres (890 mm + 160 mm).

General biology   Although the Tiger Snake is found above the snow-line where it coexists with the Copperhead *Austrelaps superba*, it is most abundant around the rivers and lakes in the lower altitudes of the Southern Highlands. It is primarily a diurnal species, although on warm summer nights it may be found on the move or basking on the road surface.

Notechis scutatus; Captains Flat N.S.W.
Photo courtesy of I. Morris

Live young are produced and these often number more than 20. The diet of the adult consists of small rodents and frogs which are common in the type of country in which the snake is found. The juveniles feed on small skinks and insects.

This thickly-built snake is normally quite docile if unmolested, however, once aroused it becomes extremely aggressive, flattening its neck and raising the anterior part of its body off the ground. Care should be exercised when handling this species as its venom is extremely toxic; only very small doses are required to kill an adult human.

Genus *Pseudechis* Wagler 1830

Type species: *Coluber porphyriacus* Shaw 1794

Head slightly distinct from the neck. Body cylindrical with smooth scales. Distinct canthus rostralis. Venom fang followed by two to five small maxillary teeth. Anal divided; anterior subcaudals entire, the remainder paired. Confined to Australia and New Guinea; absent from Tasmania.

Pseudechis porphyriacus (Shaw) **Red-bellied Black Snake**

1794 *Coluber porphyriacus* Shaw   *Zool.N.Holl.*p.27.Pl.10

Localities A.C.T.   Canberra, Cotter River, Mt Majura, Tidbinbilla Nature Reserve, Murrumbidgee River, Molonglo River

             N.S.W.   Braidwood, Yass, Gunning, Nerriga, Bungonia, Queanbeyan, Cooma, Captains Flat, Clyde Mountains, Gundaroo, Tumbarumba, Michelago

             Vic.   Murrindal, Gelantipy, Wulgulmerang, Tullochard, Dartmouth, Omeo, Strathbogie Ranges, Tintaldra

Diagnosis   The mid-body scales are smooth and arranged in 17 rows. The ventrals number between 180 and 202. The subcaudals number 54, of which the last two-thirds are divided. The anal is divided. The head is indistinct from the neck. There is a distinct canthus rostralis. The frontal is nearly as broad as it is long, but it is smaller than the supraoculars. A single nasal barely makes contact with the preocular.

Dorsally the snake is glossy black. Some specimens have a light brown snout. The ventrals and the bottom row of body scales, adjacent to the ventrals are bright red or pink. The posterior margin of each ventral and adjacent body scale is black. The subcaudals are grey. The average length of adults is 1230 millimetres (1050 mm + 180 mm).

General biology   *Pseudechis porphyriacus* is a large thick-set snake that has been known to reach a length greater than 2100 millimetres. This species, with the Common Brown Snake *Pseudonaja textilis*, would be one of the most ubiquitous species of snakes in the Southern Highlands.

As it is confined to the lower altitudes, it may be encountered in any one of a number of habitat types. It is diurnal in habit and may often be seen basking in quiet spots along rivers or other water-soaked areas where it is most abundant.

Pseudechis porphyriacus; Canberra A.C.T.

A capable swimmer, the Black Snake shows no hesitation in taking to the water as a means of escape.

The eggs are retained inside the female, and the free-born young may number more than 20. Although this species is known to be cannibalistic in captivity, it is doubtful whether snakes form a large proportion of its diet; in the wild, its diet consists mainly of frogs, lizards and small mammals.

This normally docile snake tends to flee rather than stand its ground, however, when cornered it inflates its single lung and expels the air in loud hisses. This procedure is associated with some degree of neck flattening. The venom is mildly toxic; it is predominantly haemolytic in nature and will cause rupture of blood vessels in the victim.

Genus *Pseudonaja* Günther 1858

Type species: *Pseudonaja nuchalis* Günther 1858

Body slender. Head small and not distinct from the neck. Canthus rostralis distinct. Nasal in contact with the preocular. Anal divided; all subcaudals divided. Pupil round. Venom fang followed by eight to ten maxillary teeth. Oviparous. Distributed throughout Australia and adjacent islands including Papua New Guinea; absent from Tasmania.

Pseudonaja textilis textilis (Duméril and Bibron)
Common Brown Snake

1854 *Furina textilis* Duméril and Bibron

Erpét.Gén. **7**:1242

Localities	A.C.T.	Mt Ainslie, Black Mountain, Molonglo River, Cotter River, Murrumbidgee River, outlying suburbs of Canberra
	N.S.W.	Lake George, Tarago, Bungendore, Goulburn, Gundaroo, Yass, Williamsdale, Bredbo
	Vic.	Murrindal, Gelantipy, (Buchan), Berringama, Bright, Licola, Tongio, Mt Buffalo National Park, Benambra, Strathbogie Ranges, Dartmouth

Pseudonaja textilis (juvenile banded colour phase); Beecroft Peninsula N.S.W

Pseudonaja textilis; Canberra A.C.T.
Photo courtesy of I. Morris

Diagnosis The mid-body scales are smooth in 17 rows. The anal and subcaudals are paired, the latter number between 48 and 70. The ventrals number between 190 and 230. Head small and indistinct from the neck. The diameter of the eye is greater than the distance between it and the margin of the lip. The frontal is twice as long as it is broad, being equal in width to the supraoculars. The single nasal is in contact with the preocular. The rostral is large and apparent from above. The canthus rostralis is distinct.

The colour may vary from light tan to dark brown. The ventrals are cream with a number of orange blotches which extend throughout the length of the body. The average length of adults is 1260 millimetres (1030 mm + 230 mm).

General biology This species is equally at home in dry country or in swampy land. It is sympatric with the Tiger

233

Snake *Notechis scutatus* at Lake George, where it occurs along the foreshores. The diet consists of small mammals, frogs and lizards.

This diurnal species is oviparous, producing soft-shelled eggs which often number more than 20. The eggs are deposited under rocks or well-buried logs. The young, which hatch in late summer, are characteristically light brown with a black head and nape separated by a light brown collar. The orange blotches on the ventrals are present at birth. Some juveniles have black bands which extend along the entire length of the body. The bands persist for the first year, gradually fading with age.

Brown Snakes, being lightly built are able to move extremely rapidly. When aroused this snake adopts an aggressive posture, characteristic of a number of members of the genus, raising the forebody off the ground in a series of tight S's, with its mouth agape. It may lunge periodically at its tormentor. Great care should be exercised when handling this species as it is regarded as being very dangerous to man.

Genus *Unechis* Worrell 1961

Type species: *Hoplocephalus carpentariae* Macleay 1887

Small to moderately-sized snakes; less than one metre in length. Head barely distinct from the neck. Canthus rostralis absent. Subcaudals and anal scale entire. Three to five solid maxillary teeth follow the venom fang. Internasals present. Body covered with smooth scales in 15 to 17 rows at the mid-body. Distributed throughout most regions of mainland Australia.

In 1961, Worrell split Krefft's 'mega-genus' *Denisonia* into a number of smaller genera. McDowell (1970) examined a range of characters from representatives of Worrell's genera, *Unechis*, *Suta* and *Parasuta*, concluded that all three were congeneric, and established the genus *Suta* as the senior synonym. In this book, the treatment of the species *flagellum* and *gouldii* is that adopted by Cogger (1975), namely regarding *Suta* as sufficiently distinct in possessing 19 mid-body scale rows to warrant recognition as a monotypic genus.

Key to species

Mid-body scales in 15 rows...*gouldii*
Mid-body scales in 17 rows...*flagellum*

Unechis flagellum (McCoy)　　　　**Little Whip Snake**

1878 *Hoplocephalus flagellum* McCoy
　　　　　　　Prodr.Zool.Victoria,dec.**2**.p.7.Pl.*xi*,Fig.1

Localities　A.C.T.　nil
　　　　　　　N.S.W.　Tarago, Bungendore, Captains Flat
　　　　　　　Vic.　　Strathbogie Ranges, (Myrtleford)

Diagnosis　Mid-body scales are smooth and arranged in 17 rows. The ventrals number between 132 and 138. The sub-caudals number between 26 and 36; all are single. The anal is entire. The frontal is nearly as broad as it is long, and is twice as broad as the supraoculars. The rostral is twice as broad as it is deep, being visible from above. The single nasal is in contact with the preoculars. The temporals are small and number four.

　　Dorsally this species is orange-brown in colour. The ventrals are creamy orange. There is a glossy black crown on the head which does not extend forward past the frontal or below the temporals. The N.S.W. Southern Highlands form, unlike the Victorian variety, does not have a black bar across the nasals and internasal. The leading edge of each scale is black, giving the body surface an overall reticulated pattern when inspected closely. The average length of adults is 300 millimetres (255 mm + 45 mm).

General biology　Superficially this snake resembles *U. gouldii*. *U. flagellum* appears to have a restricted distribution on the Southern Highlands, occurring in open woodlands and grass-lands. However, further collecting may reveal a much wider distribution as the habitats of the localities listed are by no means unique in the region.

　　It is nocturnal in habit and may be found under rocks on well-drained hillsides. Its diet consists of invertebrates and small skinks of the genera, *Leiolopisma* and *Hemiergis* which occur in similar microhabitats.

The Little Whip Snake is ovoviviparous. The female retains the embryos and gives birth to live young, which number two or three. The juveniles measure approximately 70 to 80 milli-metres at birth.

This species has a strange method of defending itself when handled roughly, coiling itself into a series of tight curls and flattening its body. The snake then resembles a shallow cone with its head at the apex. It is reluctant to bite when handled; the fangs are reduced and the venom dosage would be too small to cause any harmful effects to a healthy adult human.

Unechis flagellum; Bungendore N.S.W.

Unechis gouldii (Gray)   **Black-headed Snake**

1841 *Elaps gouldii* Gray   Grey — *J.Exped.Disc.Aust.*2:444

Localities A.C.T.   Coppins Crossing, Mt Ainslie, Woden
Valley, Black Mountain
N.S.W.   Goulburn, Cooma, Yass
Vic.   nil

Diagnosis Mid-body scales are smooth and arranged in 15 rows. The ventrals number between 144 and 150. The anal and subcaudals are all single, the latter number from 23 to 31. The pupil is round. The frontal is slightly longer than it is broad, and twice as broad as the supraoculars. The divided nasal is in contact with the preoculars. The supralabials number six.

The colour is light brown dorsally with cream ventrals and subcaudals. The head has a black cap, which extends down the neck and past the eyes to the upper margin of the supralabials. The preoculars may be black. The average length of adults is 335 millimetres (295 mm + 40 mm).

General biology *Unechis gouldii* is a small nocturnal species commonly found under rocks on well-drained slopes. This species is restricted to the warmer regions of the Southern Highlands, occurring in dry sclerophyll forests and woodlands.

The young, which usually number two or three, are free-born, and measure approximately 130 millimetres at birth. The diet of this species consists of small skinks and possibly insects too.

The Black-headed Snake has a wide distribution on mainland Australia, extending across the continent, even through the arid inland. In some areas, the most common coloration is as above, with an extension of the black cap in the form of a black vertebral stripe, whilst in other areas melanic forms are most common.

Unechis gouldii; Macquarie Marshes N.S.W.

Care should be observed when handling this species because, although it is a small species, it is capable of delivering quite a painful bite.

Genus *Vermicella* Günther 1859

Type species: *Calamaria annulata* Gray 1841

Head not distinct from the neck, lacking a distinct canthal ridge. Cylindrical body covered with smooth scales. Nasal in contact with preocular. Anal and subcaudals paired. Venom fang followed by three maxillary teeth. Distributed throughout mainland Australia; uncommon in the south-eastern sector.

Vermicella annulata (Gray)   **Bandy-Bandy**

1841 *Calamaria annulata* Gray

Grey — *J.Exped.Disc.Aust.***2**:443

Localities A.C.T.   Ginninderra
        N.S.W.   Yass, Queanbeyan, Tumut
        Vic.   Omeo, (Euroa)

Diagnosis   The mid-body scales are smooth and arranged in 15 rows. Ventrals number from 204 to 242. The subcaudals number from 17 to 26; all are divided. The anal is paired. The frontal is somewhat triangular in shape, being about twice as long as it is broad (at the broadest point). The nasals are single and in contact with the preoculars. There is no canthus rostralis. Internasals are present and obliquely disposed. The diameter of the eye is approximately equal to its distance from the mouth.

There is very little geographic variation in the coloration of this species; it is banded black and white down the entire length of the body. The bands encircle the whole body in an annular series. The average length of adults is 615 millimetres (570 mm + 45 mm).

General biology   The distribution of the Bandy-Bandy in the Southern Highlands is probably quite extensive, although it cannot be regarded as common.

A medium-sized species of moderate proportions, *V. annulata* is fossorial in habit and only ventures on to the soil surface at night, usually after a rainstorm. Its diet consists primarily of blind snakes (*Typhlina* spp.) and probably small fossorial skinks. *V. annulata* is oviparous and the eggs are usually deposited under rocks or logs on the ground.

When aroused this species adopts an unusual defensive attitude, arching its body into a series of loops which are held well off the ground. During this action the body is flattened considerably. The Bandy-Bandy is reluctant to bite and is not

241

Vermicella annulata
Photo courtesy of H. G. Cogger

considered dangerous to man. The fangs, and therefore the amount of venom injected, are too small to cause any serious effects to a healthy adult.

Blind snakes: family *Typhlopidae*

Blind snakes are identified as small to moderate-sized reptiles with fossorial habits. The body, which is of uniform thickness, is covered with extremely smooth, imbricate, cycloid scales. The ventral scale rows are not differentiated into wider plates as they are in all other land snakes. The snout is blunt and the reduced mouth is situated towards the ventral surface. The head is covered with large regular plates. The eyes are rudimentary, appearing as pigmented areas beneath translucent plates. The tail is short and blunt with a small ventrally-displaced terminal spine.

There are no teeth on the palate or the lower jaw. There are no movable cranial bones and the pelvis is reduced to a single bone.

Typhlopids are oviparous and lay a few large and elongate eggs.

All Australian members of the family belong to a single genus, *Typhlina*, and occur in most regions of the mainland, particularly the tropics, but can be regarded as being extralimital in the high country of south-eastern Australia.

Genus *Typhlina* Wagler 1830

Type species: *Typhlina septemstriatus* Schneider 1801

Small to moderate-sized snakes. Body more or less of uniform thickness and rounded at each end. Eyes rudimentary. Short postanal tail with a short terminal spine. Oviparous. Distributed through the Indo-Malaysian Archipelago and mainland Australia.

Morphologically this genus cannot be separated from other typhlopid genera. The only criteria for establishing this genus are the conditions of the male reproductive organs and cloacal pouches.

Typhlina nigrescens (Gray) **Blind Snake** or **Worm Snake**

1845 *Anilios nigrescens* Gray   *Cat.Liz.Brit.Mus.*p.133

Localities
A.C.T.	Mt Ainslie, Black Mountain, Coppins Crossing, Canberra
N.S.W.	Gundaroo, Yass, (Cootamundra), Lake George, Goulburn, Nimbin
Vic.	Dartmouth, Bright, Corryong, Strathbogie Ranges, Tolmie, Bandongadale, (Yea)

Diagnosis   The nasal cleft is in contact with the first labial. The mid-body scales are in 22 rows. The snout and tail are blunt. The tail region is short and terminates in a small ventrally-displaced spine. There is no differentiation of the ventral scales into plates as there is with other land snakes.

The dorsal colour is grey with a pinkish tinge. The ventrals are creamy pink. The average length of adults is 355 millimetres (340 mm + 15 mm).

General biology   *Typhlina nigrescens* is a widespread species throughout the warmer regions of the Southern Highlands, but it can not be regarded as being common to any particular locality. Despite the wide range of habitats frequented by this species, there is one factor that is common to all — *T. nigrescens* inhabits only country which is well drained and is never subjected to flooding.

The Blind Snake is a fossorial reptile whose blunt snout and very smooth homogeneous body scales represent a morphological adaptation to this mode of life. The eyes have also undergone specialisation for the Blind Snake's fossorial existence, becoming rudimentary to the degree that they appear as pigmented spots only. These snakes are frequently found on the surface of the ground at night during the warmer months of the year. They are however, more often found under well-buried rocks or logs. Its diet consists of earthworms and small soft-bodied insects and their larvae.

Typhlina nigrescens; Coppins Crossing A.C.T.
Photo courtesy of J. C. Wombey

Other than the fact that it is oviparous, little is known about the reproductive biology of this species.

When handled, *T. nigrescens* twists itself into a series of knots forming a tight and compact ball. This trait is shared in common with a large number of other members of the family. This species is one of the largest members of the genus, often attaining a length greater than 400 millimetres.

5
Snake Venoms, Snake-bite and Treatment

The nature of snake venoms

Snake venoms are mixtures of complex protein substances and enzymes which may be categorised according to their effects on animal tissues. Two main types are recognised: (i) *haemotoxins* which cause breakdown of the blood cells and of the linings of the blood vessels, with accompanying damage to surrounding and underlying tissues at the location of the bite; (ii) *neurotoxins* which attack the central nervous system and bring about death by cardiac arrest or by suffocation due to inhibition of breathing movements.

Within these two categories, the following list of components has been described by N. H. Fairley[1].

1 **Neurotoxins** (as above) acting on the bulbar and spinal ganglion cells.
2 **Haemorrhagins** destroying the endothelial cells lining the blood vessels.
3 **Thrombase** producing intravascular thrombosis.
4 **Haemolysins** destroying the red blood corpuscles.
5 **Cytolysins** acting on blood corpuscles, leucocytes and tissue cells.
6 **Antifibrins** or **anticoagulins** retarding the coagulation of the blood.
7 **Antibactericidal agents** promoting the likelihood of secondary infections.
8 **Ferments** and **kinases** for the purpose of preparing the tissues of the prey for pancreatic digestion.

[1] FAIRLEY, N. H. (1929) 'The present position of snake bite and the snake bitten in Australia.' *Med.J.Aust.*1:296–313

Of the world's venomous snakes, the viperine species (not represented in the Australian region) have the haemotoxic venoms, while the elapine snakes have the neurotoxins plus elements 4 and 5 above.

Elapine venoms are very potent and the world's most deadly species are included in this sub-family. The venom of the Mainland Tiger Snake *Notechis s. scutatus* is the most powerful per given volume of any terrestrial vertebrate and ranks amongst the most lethal of animal toxins. It is fortunate that this animal is not only rather shy but is often not capable of injecting large volumes of venom when it does bite. Needless to say, the bite of this species is always to be treated as serious.

Treatment of snake-bite

The treatment of snake-bite may be considered under two categories: (a) first aid and (b) medical treatment.

(a)   First aid

Unfortunately there are many conflicting recommendations for the immediate treatment of snake-bite. Some of these are actually dangerous — probably more dangerous than the bite itself in many instances.

First aid is not as straightforward as it might seem to be. Until quite recently, the recommended method was to lance deeply the area of the bite with a clean razor blade or knife, and to apply suction by mouth[1] to the wound. The chances of secondary complications arising from the application of this treatment are considerable and fortunately the method has fallen into disrepute.

Until quite recently another accepted first aid treatment was to apply an arterial tourniquet. However, experimental studies have revealed that the central spread of venom from

[1] Snake venom is not toxic when ingested, although considerable danger exists if there are any lesions in the mouth or oesophagus, such as ulcers or bad teeth.

the bite occurs mainly by way of the lymphatic system and to a lesser degree in the blood venous system. This fact, coupled with acute dangers inherent in the application of a tourniquet by untrained persons, has led to the adoption of the following recommendations for first aid treatment:

1. *Apply a broad, comfortable, constrictive bandage **(not an arterial tourniquet)** to thigh or upper arm as appropriate* [most bites occur on the limbs, rarely on face or trunk]. *The bandage should have the same tension as a crepe bandage applied to a sprained ankle.*
2. *Keep the victim and the affected limb as still as possible; treat as for fracture, elevate the limb, use a splint or arm sling.*
3. *Since the most likely immediate symptoms are fright and fear of death, convincing reassurance is vital at all times. Death from snake bite is rare* [see page 252], *and several hours may elapse in the absence of medical treatment before the victim's condition becomes critical.*
4. *Whenever possible bring transport to the victim rather than vice versa.*
5. *If possible, notify the hospital of the impending arrival of the victim.*
6. *Leave the constrictive bandage in place until the victim has reached medical care.* [Premature release of the bandage may result in sudden systemic envenomation].

In addition to this sound advice we add the following notes.

A. It is often inadvisable to try and kill the snake which has been responsible for the bite as it is probably still capable of inflicting further effective bites (see below). It is certainly better to be able to identify the animal accurately in the field and so aid in the ultimate choice of antivenene.

B. Most Australian snakes have to hold and 'chew' on the bite in order to work the venom into the wound, it is quite likely that the bite itself has not received much (if any) dose, and most of the venom will be on the surface of the skin. This is most likely to be the case when the bite has been inflicted through clothing. However, contrary to earlier advice, do not wash the wound as this may result in the transfer of more surface venom into the bite.

C. The fang punctures may not be visible to the naked eye (in some cases these have not been revealed until *post mortem*). Even if they are visible, they may not necessarily be represented by a 'classical' pattern of punctures and may appear as a few light scratches. Multiple, random fang marks may indicate massive envenomation has occurred.

D. Remember that few doctors have had practical experience in the treatment of snake-bite, and probably have never encountered the vital early stages immediately following the attack. By the time the victim has reached the surgery or hospital, he may be suffering more from shock and amateurish first aid treatment than from the direct effects of the bite itself. Be ready to recount exactly what happened at the time of the bite, when it occurred, and what subsequent first aid steps were taken.

From this point on it is up to professional medical personnel to deal with the patient.

(b)   Medical treatment

Modern treatment of snake-bite involves the use of *specific antivenenes*. These are a number of different sera, each of which is used to counter the effects of the bite of one particular species of snake. The following is an outline of how specific antivenenes are formulated and produced.

There must be a good supply of the chosen species of snake. Snakes used in the production of antivenenes may either be kept on snake-farms, or the serum laboratory may depend on amateur or professional snake catchers. The next step is to *milk* the snakes of their venom. This is done by causing the snake to puncture with its fangs a diaphragm stretched across the mouth of a vial. Holding the snake firmly behind the angle of the jaws, the jaws are worked over the neck of the vial. As soon as the diaphragm is punctured, the venom is expressed by massaging the venom glands in the snake's head.

When sufficient venom has been collected in this manner, a number of suitable detoxicant chemicals are added and the mixture is inoculated into a horse. The horse builds up an immune response to the venom proteins and, after a suitable period, a serum containing the appropriate antibodies is prepared from the horse's blood.

Antivenenes are certainly the most efficient treatment for snake-bite discovered so far, but their use is by no means a cure-all. Some people are allergic to the horse serum itself and may develop a serious reaction to it following its administration.

Most antivenenes are specific and it is for this reason that it is important to identify the snake responsible for inflicting the bite. There has been some encouraging research into the production of polyvalent antivenenes which would be active in the treatment of the bites of a number of species of snake, but because of the varied composition of the different venoms it is usually better to deal with each case specifically. The use of polyvalent antivenenes may be restricted to cases of doubtful identity.

Incidence of snake-bite in the Southern Highlands region

Taking into account the number of large venomous species in the Southern Highlands region and their relative abundance, the statistics of incidence of snake-bite are reassuring. Unfortunately, a survey by us of regional hospitals yielded little in the way of useful data. The inaccessibility of relevant data is due to the official mode of classification of snake-bites within hospital records. The International Classification of Disease Coding Index System (I.C.D.) has *Code E905*; *Bites or stings by venomous insects or animals* under which the bites of spiders, snakes, venomous sea animals, centipedes and stings of bee, scorpion, wasp or insect [*sic.*] are all grouped; it is the larger hospitals, and hence those most likely to receive snake-bite victims, which use this particular coding. However,

251

in the last ten years, only three snake-bite cases were reported by two of the district hospitals in the region. One bite was of the Brown Snake *Pseudonaja t. textilis*, another the Tiger Snake *Notechis s. scutatus*, and the third unspecified. In addition to these we can report from personal knowledge (RWGJ from experience!) the treatment of four more cases of bites involving the above two species. In each instance, the victim was handling the snake.

One hospital in our survey reported that a good deal of antivenene had been lent to veterinary surgeons for treatment of horses suffering from snake-bite.

The urge to kill snakes is widespread but we believe that this should be discouraged for two reasons. Firstly, the assailant places himself in severe danger of being bitten and secondly, it must be remembered that snakes play an important ecological role as predators. In addition, many species which are harmless to man together with snake 'look-alikes', such as pygopodid lizards and others, often succumb to these senseless attacks. This is demonstrated by the large number of reptile specimens, mangled beyond recognition, which have been brought to us. One unfortunate animal which is particularly prone to this fate is the Bluetongue *Tiliqua s. scincoides* whose sinuous mode of progression and short limbs, which are often lost to view in long vegetation, render it all too snake-like.

Common sense 'Do's' and 'Dont's'

If you want to avoid being bitten:

★ **DON'T** attempt to kill a large snake; its ability to move fast and with accuracy is considerably greater than yours.

★ **DON'T** think that a spade or a shovel is a safe weapon with which to attack a snake; many species are sufficiently agile to be able to take avoiding action and make an effective follow-up strike.

★ **DON'T** run through long grass or other vegetation.

- ★ **DON'T** step over or off a rock or a log; a snake can easily be hiding in its lea.
- ★ **DO** take great care in the bush in the early morning when the sun is beginning to warm the ground; snakes are sluggish and disinclined to move away under these conditions.
- ★ **DON'T** ever tease or torment a snake in an effort to make it respond; the result may be more than you bargained for.
- ★ **DO** clear rubbish, wood piles and other likely snake refuges away from dwellings, outhouses and places where children and pets play.
- ★ **DON'T** roll objects away from you so that the cavity beneath is exposed towards you, if you are clearing debris or moving rocks or logs.
- ★ **DO** wear stout gloves and equip yourself with a crowbar or baling hook to aid in moving heavier objects.
- ★ **DO** wear stout clothing when hiking in the bush; denim jeans tucked into thick socks, although not completely snake-proof, can cut down on the effectiveness of a bite.
- ★ **DON'T** reach into a hollow log or other such cavity without making sure that it doesn't harbour a snake.
- ★ **DON'T** handle a snake unless (a) you **know** it is harmless; (b) you can accurately identify it as harmless; or (c) you know how to handle it properly; remember, many expert snake handlers get bitten.

If you do handle snakes:

- ★ **DON'T** get over confident or become exhibitionist; spectacular bites occur in front of audiences.
- ★ **DO** make sure that the nearest doctor or hospital has current stocks of antivenene for the species of snake in your possession.
- ★ **DON'T** handle highly venomous snakes if no-one else is present.

Abbreviations of scientific publications

Ann.Mag.Nat.Hist.	*Annals and Magazine of Natural History.* London
Ann.Nat.Hist.	*Annals of Natural History.* London
Ann.Mus.d'Hist.Nat.Paris	*Annales du Museum d'Histoire Naturelle.* Paris
Ann.Philos	*Annals of Philosophy.* London
Arch.Naturgesch.	*Archiv für Naturgeschichte.* Berlin
Aust.J.Zool	*Australian Journal of Zoology.* Melbourne
Cat.Colub.Sn.Brit.Mus.	*Catalogue of Colubrine Snakes in the Collection of the British Museum (Natural History).* London
Cat.Liz.Brit.Mus.	*Catalogue of Lizards in the Collection of the British Museum (Natural History).* London
Erpét.Gén.	*Erpétologie Générale, ou Histoire Naturelle Complète des Reptiles.* Paris
Gen.Zool.	*General Zoology.* London
Grey — *J.Exped.Disc.Aust.1837–39*	George Grey — *Journals of two expeditions of discovery in north western and Western Australia during the years 1837, 38, 39 . . . with observations on the moral and physical conditions of the aboriginal inhabitants.* Boone, London (Reprinted 2 vols)
Herpetofauna	*Journal of the Australian Herpetological Society.* Sydney
J.Proc.R.Soc.West Aust.	*Journal and Proceedings of the Royal Society of Western Australia.* Perth
J.Zool.London	*Journal of Zoology.* London
K.Sven.Vetenskapsakad.Handl.	*Kungliga Svenska Vetenskapsakademiens Handlingar.* Uppsala and Stockholm
Med.J.Aust.	*Medical Journal of Australia.* Sydney
Mem.Natl.Mus.Victoria,Melbourne	*Memoirs of the National Museum of Victoria.* Melbourne
Misc.Publ.Mus.Zool.Univ.Mich.	*Miscellaneous Publications of the Museum of Zoology, University of Michigan.* Ann Arbor
Monatsber.K.Preuss.Akad.Wiss.Berlin	*Monatsberichte der Königlichen Preussischen Academie der Wissenschaften zu Berlin*
Nat.Miscell.	*Naturalists Miscellany.* London
Phys.Serp.	*Essai sur la Physionomie des Serprens. Partie Descriptive.* Amsterdam

Proc.Linn.Soc.N.S.W.	*Proceedings of the Linnean Society of New South Wales.* Sydney
Proc.R.Soc.Victoria	*Proceedings of the Royal Society of Victoria.* Melbourne
Proc.R.Zool.Soc.	*Proceedings of the Zoological Society.* London
Proc.Zool.Soc.Victoria	*Proceedings of the Zoological Society of Victoria.* Melbourne
Prodr.Zool.Victoria	*Prodromus of the Zoology of Victoria.* Melbourne
Rec.Aust.Mus.	*Record of the Australian Museum.* Sydney
Rec.S.Aust.Mus.	*Record of the South Australian Museum.* Adelaide
Règne Anim.	*Règne Animal.* Paris
Senkenbergiana Biol.	*Senkenbergiana Biologica. Senkenbergische naturforschende Gessellschaft.* Frankfurt a.M.
Sitzungsber.Ges.Naturforsch.Freunde Berlin	*Sitzungsberichte der Gessellschaft Naturforschenden Freunde zu Berlin*
*Voy.*Uranie.*Zool.*	*Voyage Autour du Monde de la Corvette* l'Uranie *Zoologie*
West.Aust.Nat.	*Western Australian Naturalist.* Perth
White — *J.Voy.N.S.W.*	John White — *Journal of a voyage to New South Wales with 65 plates of nondescript animals, birds, lizards, serpents, curious cones of trees and other natural productions.* Bebrett, London
*Zool.Voy.*Erebus *and* Terror *Rept.*	*The zoology of the voyage of H.M.S.* Erebus *and* Terror *under the command of Captain Sir James Clark Ross, R.N., F.R.S., during the years 1839 to 1843. Reptiles, fishes, Crustacea, insects, Mollusca.* E. W. Jansen, London
Zool.N.Holl.	*Zoology of New Holland.* J. Sowerby, London

Glossary

Adaptive radiation Evolution, from a more primitive type of organism, of several divergent forms adapted to distinct modes of life, e.g. in the Mesozoic era (*c.* 180 million years ago) the basal stock of the reptiles radiated into many forms adapted to running, flying, swimming, burrowing, etc.

Adpress To press closely against the body, e.g. adpressed limbs.

Allopatric Having separate and mutually exclusive areas of geographical distribution. See **Sympatric**.

Allotype Paratype of the sex opposite to that of the holotype.

Anal The terminal plate in the ventral scale series in snakes, often larger than the ventral scales and free along most of the lateral margin. May be single or divided by a diagonal suture along the midline. Pertaining to the anus.

Annulate Furnished with, or marked with, rings (from Latin *annulatus, annulata*).

Anterior Situated at, or relatively nearer to, the front, e.g. the situation of the head.

Apodous Without legs.

Appendage Any considerable projection from the body of any animal, e.g. leg or tail.

Aquatic Pertaining to, confined to, or living in, water. The *aquatic* environment. See **Terrestrial**.

Arboreal Pertaining to, living on, or among, trees.

Arthropod A member of the largest animal phylum, the *Arthropoda*, which includes crabs, insects, spiders, scorpions, centipedes. These animals have a hard, jointed exoskeleton and paired, jointed legs.

Auditory organ Sense organ for detecting sound; the ear.

Auricular lobule In lizards, large or otherwise-modified scales projecting over the ear opening anteriorly.

Autotomy Spontaneous or reflexive separation of a part

of an organism, e.g. the tail, from the organism across a vertebral plane of weakness (fracture plane) by convulsive contractions of the muscles. *Syn.* taildropping.

Azygous Not one of a pair; single. Applied herpetologically to the condition of some scales.

Bifid Divided into two lobes by a median cleft. *Syn.* bifurcated.

Biology The study of living organisms.

Buccal Pertaining to the mouth; buccal cavity, the interior of the mouth.

Calciferous Containing calcium, e.g. calciferous eggshell.

Canthus ridge A sharply angulate canthus rostralis; clearly delimited border line between the side of the snout and the top of the head.

Canthus rostralis The angle of the head from the tip of the snout to the anterior end of the eyebrow which separates the dorsum of the head from the side of the snout and the loreal region.

Carapace Shield of the exoskeleton covering part of the body. The dorsal part of the shell of *Chelonia*, consisting of exoskeleton plates fused with the ribs and vertebral column. See **Plastron**.

Carinate Possessing keels or ridges, keeled, also *carina*, keel (q.v.).

Carnivore Flesh-eater. See **Herbivore, Omnivore**.

Caudal Pertaining to the tail. *Caudal* scale; any scale or plate on the tail of a reptile.

Cervical Pertaining to the neck.

Character A morphological attribute pertaining to a species. Used in the process of classification for grouping animals within taxa.

Chin-shield (a) any of the paired, elongated scales on the lower jaw of snakes between the lower ends of the *labials*.

(b) the paired series of scales on the mid-line of the lower jaw posterior to the *mental* scale in lizards.

Ciliary Pertaining to, or associated with, the eye,

eyelid or orbit. Frequently used with the prefix *super, supra,* or *infra* to signify the position of scales in relation to the eye itself.

Climax
Type of plant community, the composition of which is more or less stable, in equilibrium with existing natural environmental conditions, e.g. dry sclerophyll forest.

Cline
Continuous gradation of form differences in a population of a species, correlated with its geographical or ecological distribution. Hence, *clinal* variation.

Cloaca
The common chamber into which the intestinal, urinary and reproductive ducts discharge their products, opening to the outside through a single vent.

Community
Ecological term for any naturally occurring group of different organisms inhabiting a common environment, interacting with each other (especially through food relationships) and relatively independent of other groups. Communities may be of variable size and larger ones may contain smaller ones. Also assemblage, association, consociation.

Compressed
Pressed together, flattened; strongly ovate in cross section; laterally compressed = vertically flattened; dorso-laterally compressed = horizontally flattened.

Congeneric
Belonging to the same genus.

Conspecific
Belonging to the same species.

Cotype
See **Syntype**.

Crepuscular
Active or appearing at dusk or dawn.

Dactyl
Pertaining to the toes or fingers.

Didactyl
Having two digits on a foot.

Digit
A finger or toe.

Digital
Pertaining to the digits. *Digital* scale, a scale found on the digit of any foot, between the palm or sole and the claw.

Diurnal
Active or appearing in the daytime.

Dominant
Of a plant species, the most characteristic and common species in a particular plant community to a large extent governing the

258

	type and abundance of other species in the community.
Dorsal	Situated on, or relatively nearer to, the back. The side of the animal that is usually directed upwards with reference to gravity. See **Ventral**.
Dorso-ventral	The plane bisecting the dorsal and ventral surfaces.
Dorso-lateral	The uppermost part of the sides of an animal, the upper flanks.
Ecology	Study of the relationships of animals and plants, particularly of animal and plant communities to their surroundings, animate or inanimate.
Ectotherm	An organism whose body temperature approximately follows that of its surroundings, hence *ectothermic*. *Syn.* cold-blooded, poikilotherm. See **Endotherm**.
Endemic	Confined to a given region.
Endotherm	An organism that regulates body temperature by means of internal, metabolic processes, with well-developed homeostatic mechanisms for so doing. Not true of reptiles and amphibians. *Syn.* warm-blooded, homeotherm. See **Ectotherm**.
Environment	Collective term for the conditions in which an organism lives, e.g. temperature, light, air/water, other organisms with which it interacts, etc.
Evolution	Cumulative change in the characteristics of populations of organisms occurring in the course of successive generations related by descent.
Fauna	The animal population present in a certain locality, natural community, region, country, etc. *Faunistic*, relating to the fauna.
Femoral	Pertaining to the thigh region of the leg.
Femoral pore	An opening in the centre of a single, usually enlarged scale on the ventral surface of the thigh in many lizards. The function is unknown.

Filiform	Resembling a filament; threadlike.
Flora	The plant population present in a certain locality, natural community, region, country, etc. *Floristic*, pertaining to the flora. *Floral*, pertaining to a flower.
Form	One of the kinds of polymorphic species; a seasonal variant; or used as a neutral term in taxonomy when it is unclear whether a species or sub-species or some minor grouping is the appropriate classification.
Fossorial	Pertaining to animals which dig dwellings and seek their food in the soil; adapted for digging or burrowing.
Frontal	(a) the large median, unpaired plate on top of the head between the eyes in snakes and most lizards. (b) one of the component bones of the skull.
Frontonasal	A cephalic scale or scales in reptiles located between the internasals, prefrontals and loreals. (See Figures 5, 6, 7, 10.)
Frontoparietal	Plate or plates on the dorsum of the head between the frontals and parietals in lizards. (See Figure 5.)
Girdle	Skeletal support in the body wall for the attachment of limbs: e.g. shoulder girdle (pectoral girdle) for the attachment of the forelimbs, and hip girdle (pelvic girdle) for the attachment of the hindlimbs.
Granular scale	A very small flat scale, distinguishable only with the aid of strong magnification.
Gular	Pertaining to the throat region. Gular fold, a transverse fold of skin of the throat immediately anterior to the insertion of the forelimbs.
Habitat	Place with a particular kind of environment inhabited by organism(s) e.g. swamp, grassland, fissures in logs, crevices in rocks, etc.
Heliotherm	An ectothermic animal which utilises the sun's radiation as a primary source of heat in thermal regulation. See **Thigmotherm**.

Hemipenis	Either one of the paired, male copulatory organs which lie laterally in a cavity in the base of the tail. Plural: *hemipenes*.
Herbivore	Plant-eater. See **Carnivore, Omnivore**.
Herpetology	The study of reptiles and amphibians; the natural history of reptiles and amphibians.
Holotype	The single specimen chosen for the designation of a new species and upon which its published description is drawn. Ideally the holotype should be the most representative specimen selected from a series. The holotype is deposited in a museum collection and all scientific comparisons of other individuals of the species are made with it.
Imbricate	Arrange, to be arranged, so as to overlap like the tiles on a roof. *Imbricate* scales, body scales of reptiles where the posterior edge of one scale overlaps the anterior edge of the following scale.
Immaculate	Spotless, unpigmented, unmarked.
Internasals	The plates on the dorsum of the head in lizards and snakes, lying between the nasals.
Interparietal	(a) in lizards, a scale on the dorsal mid-line of the head, lying between the parietals. (See Figure 5.) (b) in blind-snakes, may be used for any of the mid-line scales on the dorsum of the head. (See Figure 7.)
Invertebrate	Collective term for all animals which do not possess a spinal column, e.g. insects, spiders, molluscs, annelids (earthworms), etc. See **Vertebrate**.
Juxtaposed	Lying side-by-side, not overlapping; used in herpetology to refer to the character of the scales. See **Imbricate**.
Keel	A ridge-like process. In reptiles, the slight ridge in the cutaneous part of an individual scale in some species. Also carina.
Labial	Any one of the row of scales that borders the lip in reptiles; of the upper lip, *supralabial*; of the lower lip, *infralabial, sublabial*.

Lamella	Any one of a group of soft, overlapping plates, lying in a ranked series. Most often used in herpetology to describe the transverse plates on the underside of the digits in many lizards; hence subdigital *lamellae*.
Lateral	Situated at, or very near to the sides of an animal; used in herpetology with reference to a scale on the side of the body in contrast with the ventral and dorsal scales.
Lectoallotype	A specimen of the opposite sex to that of the lectotype and subsequently chosen from the original series of the collected material.
Lectotype	A specimen chosen from syntypes to designate the type of the species.
Lobule	See **Auricular lobule**.
Loreal	A plate lying on the side of the head between the nasals and preoculars. (See Figure 5.) The number may vary from one to 20 in lizards. Absent in elapine snakes.
Median	Situated on, or towards the plane which divides a bilaterally symmetrical animal or organ into right and left halves; the midline.
Melanic	Showing dark brown to black pigmentation.
Mental	The azygous scale at the anterior edge of the lower jaw in snakes and lizards, bordered on both sides by the first infralabials.
Monodactyl	Having only a single digit on a foot.
Monotypic	Of a genus which includes only one species, hence *monotypy*.
Morphology	The study of form, hence *morphological*.
Nasal	(a) a plate lying on the side of the head, recognised by the presence of the external nostril in most reptiles. (See Figure 5.)
	(b) a component dermal bone of the skull.
Nasal cleft	A deep sulcus or groove running from the lipline across the nasal scale, through the nostril to the prefrontal scale, in snakes of the family *Typhlopidae. Syn.* nasal groove.
Niche	The place or status of an organism in the environment.

Nictitating membrane	A thin transparent fold of conjunctival tissue in the inner angle of the eye, which can be drawn across the eyeball from side-to-side. *Syn.* third eyelid.
Nocturnal	Occurring at night; active by night.
Nuchal	(a) the lamina of the turtle carapace lying on the dorsal mid-line at the anterior end. (b) used in some lizards for the scales on the neck immediately behind the head. (See Figure 5.)
Ocellated	Possessing ocelli or eye-like spots.
Olfactory	Concerned with the sense of smell. *Olfactory* sense; *olfactory* organ, the nose.
Omnivore	An animal whose diet encompasses a wide range of foodstuffs, including flesh and vegetable matter.
Ophiophagy	The inclusion of snakes in the diet of any animal, including other snakes.
Orbit	The region of the eye.
Oviparous	Producing young by means of eggs which are released from the ovary, provided with membranes and/or shell by the lining of the oviduct and almost immediately expelled from the body. The entire embryonic development takes place outside the female. See **Ovoviviparous, Viviparous**.
Ovoviviparous	Producing young by means of eggs which are provided with membranes after their release from the ovary. The eggs are retained, usually in the oviducts and much of the embryonic development takes place within the body of the female. See **Oviparous, Viviparous**.
Palpebral	The minute scales covering the eyelid proper in lizards.
Palpebral disc	The transparent area (window) in the centre of the lower eyelid in certain lizards, e.g. species of the genus *Leilopisma*.
Paralectotype	A specimen of a series used to designate a species which is later designated as a *paratype*.

Paratype	A specimen described at the same time as the one described as the *type* of a new species.
Parietal	(a) either of a pair of dermal bones in the roof of the skull.
	(b) either of a pair of scales on the dorsum of the head of a snake immediately behind the frontal. (See Figure 6.)
	(c) variously used for a head scale of blind snakes. (See Figure 7.)
	(d) large scales on the dorsum of the head of lizards, either fused or paired. (See Figures 5 and 10.)
Pectoral girdle	Girdle (q.v.). Skeletal support in the body wall providing articulation for the forelimbs (shoulder girdle). Absent in all snakes.
Pelvic girdle	Girdle (q.v.). Skeletal support in the body wall providing articulation for the hindlimbs (hip girdle). Absent in all snakes except members of the family *Boidae*.
Pentadactyl	Having five fingers or toes.
Phylogeny	Evolutionary history, hence *phylogenetic*.
Physiology	The study of the normal bodily functions of living organisms.
Pineal eye	A sensory structure capable of light reception, located on the dorsal side of the brain and opening to the outside through the parietal foramen on the interparietal scale.
Plastron	The ventral part of the shell of tortoises and turtles, consisting of a series of paired bones which are overlain by a series of horny plates alternating with the bones.
Poikilotherm	See **Ectotherm**.
Population	All the members of a species existing in a locality, natural community, region, etc. amongst whom there is a regular gene flow through mating.
Posterior	Situated at, or relatively nearer to the hind end of an animal; usually refers to the end directed backwards when the animal is in motion.
Postnasal	A scale or scales which lie behind the nasal and anterior to the loreal.

Postocular	A scale or scales bounding the orbit posteriorly.
Preanal pore	In some lizards the femoral pore series may extend on to the body anterior to the anus. This term is used to distinguish those on the body from those on the leg, but all appear to be identical in nature. In some species only the preanal pores are present.
Prefrontal	(a) a scale or scales on the dorsum of the head immediately anterior to the frontal. See Figures 5, 6, 7 and 10.) (b) a component dermal bone of the skull, found on the anterior margin of the orbit.
Preocular	A scale or scales bounding the orbit anteriorly. (See Figures 5, 6, 7 and 10.)
Punctiform	Having a dot-like appearance
Radiation	See **Adaptive radiation**.
Relict (or Relic)	Localised remains of an originally much wider distribution, e.g. small areas of dry sclerophyll vegetation remaining in pine plantations.
Reticulated	Having a network or laced pattern.
Rhomboidal	Diamond-shaped.
Rostral	The scale at the tip of the snout, bordering the mouth and separating the supralabials. (See Figures 5, 6, 7 and 10.)
Rugose	Wrinkled, folded; meaning rough-surfaced with reference to the condition of the scales.
Saxatile	Living, growing, on or among rocks (from Latin *saxatilis*).
Specific	Peculiar to, or characteristic of, a species.
Spinose	Bearing many spines; usually with reference to the condition of the scales.
Striate(d)	(a) scales which are grooved with minute lines (*striae*); scales with irregular planes of elevation which lie parallel. (b) streaked.
Subcaudal	A scale lying on the ventral surface of the tail. In snakes, they may be divided, single, or both. (See Figure 8.)

Subocular	Restricted to one of more scales lying immediately ventral to the orbit. These scales, in many reptiles, include the supralabials lying adjacent to the orbit.
Supraciliary	Any one of a series of scales lying along the outer edge of the supraoculars in lizards. Quite small, usually numerous and bordering the orbit.
Supralabial	Any one of the scales bordering the upper lip in reptiles, separated anteriorly by the rostral. (See Figures 5, 6 and 10.)
Supranasal	Usually reserved for a scale or scales lying immediately above the nasal and lateral to the internasal.
Supraocular	(a) either of a pair of shields that lie above the eye in snakes. (See Figures 6 and 7.) (b) the scales lying on the dorsum of the orbit in lizards — usually enlarged, but may be the same size as other head scales. (See Figure 5 and 10.)
Suture	(a) the boundary between two abutting plates or scales. (b) the line of interlock between two saw-toothed bones, as in the skull.
Sympatric	Two or more species having the same geographical distribution, or where two distributions of species overlap. See **Allopatric**.
Syntype	Any one specimen of a series used to designate a species when the holotype and paratype have not been selected.
Temporal	(a) squamosal bone of the skull. (b) scale or scales behind the postoculars, below the parietals and above the labials, in snakes and lizards. Usually arranged in two or more vertical rows: anterior or primary row; posterior or secondary row; etc. A temporal formula of 1 + 4 indicates one primary and four secondary temporals.
Terrestrial	Pertaining to, confined to or living, on land. See **Aquatic**.

Territory
An area of the habitat occupied by an individual or group, trespassers upon which, if belonging to the same species, are attacked and driven off. Particularly, though not exclusively, concerned with breeding behaviour, a male often holding a territory alone at first and subsequently being joined by female(s), hence *territorial* behaviour.

Thigmotherm
An ectothermic animal which uses conduction of heat from the surrounding environment, such as air and soil, as a means of gaining heat during temperature regulation, but not from solar radiation. See **Heliotherm**.

Topotype
A specimen of the same species collected in the same locality (type locality) as the type specimen.

Tricarinate
A single scale or lamina with three separate and distinct keels on its surface.

Tubercle
A small, rounded, discrete hump or bump on the skin. Used in herpetology with reference to the nature of certain scales in lizards.

Tympanum
The membrane covering the external opening of the middle ear chamber or vestibule.

Vegetarian
An animal which subsists exclusively on plant food. *Syn.* herbivore. See **Carnivore**.

Vent
The cloacal opening. See **Cloaca**.

Ventral
Lower surface; any one of the large plates on the lower surface or belly of a snake, between the head and the anal plate.

Vermiform
Worm-shaped or having a worm-like habit.

Vertebral
(a) in snakes and lizards, any one of the row of scales lying on the mid-dorsal line of the body.
(b) pertaining to the backbone or mid-dorsal line.

Vertebrate
Animal possessing a spinal column. See **Invertebrate.**

Viviparous
Having embryos which develop within the maternal organism and derive nutrients by close association with the maternal circulatory system, frequently by placental attach-

ment without the interposition of any egg membranes, hence *viviparity*. See **Ovoviviparous, Oviparous**.

Further reading

BESTE, H. (1970) 'Reptiles of the Hattah District, Victoria.' *Vict.Nat.* **87**:262–65

BOULENGER, G. A. (1896) *Catalogue of the snakes in the British Museum (Natural History)* **3 & 4**. London

——————— (1896) *Catalogue of the lizards in the British Museum (Natural History)* **1 & 2**. London

BUSTARD, H. R. (1968) 'The reptiles of Merriwindi State Forest, Pilliga West, northern New South Wales.' *Herpetologica* **24**:131–40

——————— (1970) *Australian lizards.* Collins, Sydney

CHILD, J. (1969) *Australian alpine life.* Lansdowne, Periwinkle Series, Melbourne

CLARKE, C. J. (1965) 'A comparison between some Australian five-fingered lizards of the genus *Leiolopisma* Duméril & Bibron (Lacertilia, Scincidae).' *Aust.J.Zool.* **13**:577–92

COGGER, H. G. (1960) 'The ecology, morphology, distribution and speciation of a new species and subspecies of the genus *Egernia* (Lacertilia, Scincidae).' *Rec.Aust.Mus.* **25**:95–105

——————— (1967) *Australian reptiles in colour.* A. H. & A. W. Reed, Sydney

——————— (1975) *Reptiles & Amphibians of Australia.* A. H. & A. W. Reed, Sydney

COPLAND, S. J. (1945) 'Geographic variation in the lizard *Hemiergis decresiensis* (Fitzinger).' *Proc.Linn.Soc.N.S.W.* **70**:62–92

DAVEY, K. (1970) *Australian lizards.* Lansdowne, Periwinkle Series, Melbourne

FAIRLEY, N. H. (1929) 'The present position of snake bite and the snake bitten in Australia.' *Med.J.Aust.* **1**:296–313

GLAUERT, L. (1961) *A handbook of the lizards of Western Australia.* Handbook No. 6; *West.Aust.Nat.*, Perth

——————— (1967) *A handbook of the snakes of Western Australia.* 3rd edn. *West.Aust.Nat.*, Perth

GOODE, J. (1967) *Freshwater tortoises of Australia and New Guinea.* Landsdowne, Melbourne

GREER, A. E. (1967) 'A new generic arrangement for some Australian scincid lizards.' *Breviora* **267**:1–19

KINGHORN, J. R. (1956) *The snakes of Australia.* 2nd edn. Angus & Robertson, Sydney

McDOWELL, S. B. (1967) '*Aspidomorphus*, a genus of New Guinea

snakes of the family Elapidae, with notes on related genera.'
*J.Zool.Lond.***151**:497–543

—————— (1970) 'On the status and relationships of the Solomon
Island elapid snakes.' *J.Zool.Lond.***161**:145–90

MITCHELL, F. J. (1948) 'A revision of the lacertilian genus *Tym-panocryptis.' Rec.S.Aust.Mus.***9**:57–86

—————— (1950) 'The scincid genera *Egernia* and *Tiliqua* (Lacertilia).'
*Rec.S.Aust.Mus.***9**:275–308

—————— (1951) 'The South Australian reptile fauna.' *Rec.S.Aust.
Mus.***9**:545–57

—————— (1953) 'A brief revision of the four-fingered members of
the genus *Leiolopisma* (Lacertilia).' *Rec.S.Aust.Mus.***11**:75–90

MORRIS, R. & D. (1968) *Men and snakes.* Sphere Books, London
(*paperback*)

RAWLINSON, P. A. (1966) 'Reptiles of the Victorian mallee.'
*Proc.R.Soc.Victoria.***79**:605–19

—————— (1969) 'The reptiles of East Gippsland.' *Proc.R.Soc.
Victoria.***32**:113–28

—————— (1971) 'The reptiles of West Gippsland.' *Proc.R.Soc.
Victoria.***84**:37–51

—————— (1971) 'Reptiles of Victoria.' *Victorian Year Book* **85**

STORR, G. M. (1964) 'The agamid lizards of the genus *Tympanocryp-tis* in Western Australia.' *J.Proc.R.Soc.West.Aust.***47**:43–50

—————— (1964) 'Some aspects of the geography of Australian rep-tiles.' *Senckenberg.Biol.***45**:577–89

—————— (1964) '*Ctenotus*, a new generic name for a group of Aus-tralian skinks.' *West.Aust.Nat.***9**:84–8

—————— (1967) 'The genus *Vermicella* (Serpentes, Elapidae) in West-ern Australia and the Northern Territory.' *J.Proc.R.Soc.
West.Aust.***50**:80–92

—————— (1968) 'The genus *Ctenotus* (Lacertilia, Scincidae) in the
eastern division of Western Australia.' *J.Proc.R.Soc.
West.Aust.***51**:97–109

—————— (1968) 'Revision of the *Egernia whitei* species group (Lacer-tilia, Scincidae).' *J.Proc.R.Soc.West.Aust.***51**:51–62

—————— (1969) 'The genus *Ctenotus* (Lacertilia, Scincidae) in the
Northern Territory.' *J.Proc.R.Soc.West.Aust.***52**:97–108

—————— (1971) 'The genus *Ctenotus* (Lacertilia, Scincidae) in South
Australia.' *Rec.S.Aust.Mus.***16**:1–15

—————— (1972) 'The genus *Morethia* (Lacertilia, Scincidae) in West-ern Australia.' *J.Proc.R.Soc.West.Aust.***55**:73–9

WAITE, E. R. (1929) *The reptiles and amphibians of South Australia.* 270 pp. Harrison Wier, Government Printer, Adelaide

WORRELL, E. (1960) *Dangerous snakes of Australia.* Angus & Robertson, Sydney

──────── (1961) 'Herpetological name changes.' *West.Aust. Nat.***8**:18–27

──────── (1963) *Reptiles of Australia.* Angus & Robertson, Sydney

General Index

Index to scientific names

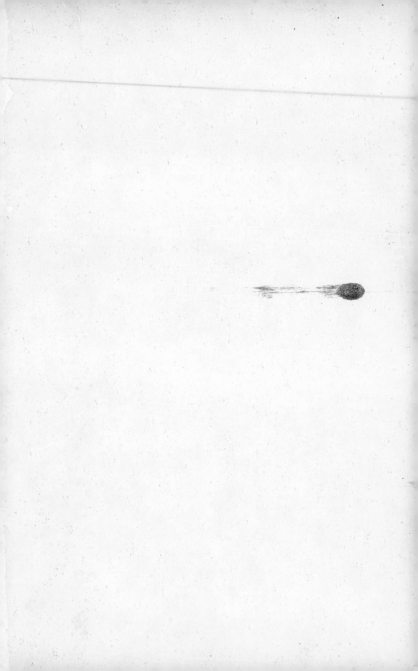